Dark web

A Complete Guide to the Dark Web
True Tales

*(The Addictive Input down able Gripping Crime
Thriller of Mystery and Suspense)*

Theodore Guerrero

Published By **John Kembrey**

Theodore Guerrero

All Rights Reserved

Dark Web: A Complete Guide to the Dark Web
True Tales (The Addictive Input down able
Gripping Crime Thriller of Mystery and Suspense)

ISBN   978-1-7382986-4-8

Legal & Disclaimer

Upon using the information contained in this book, you agree to hold harmless the Author from and against any damages, costs, and expenses, including any legal fees potentially resulting from the application of any of the information provided by this guide. This disclaimer applies to any damages or injury caused by the use and application, whether directly or indirectly, of any advice or information presented, whether for breach of contract, tort, negligence, personal injury, criminal intent, or under any other cause of action.

You agree to accept all risks of using the information presented inside this book. You need to consult a professional medical practitioner in order to ensure you are both able and healthy enough to participate in this program.

Table Of Contents

Chapter 1: What Is The Deep Web?

So yet again men and welcome over again to our extremely good Book about the complete advent to the Deep Web. So in contemporary-day beauty we're going to talk approximately what's the deep net. So we are going to see what the deep net is and what the difference is the various deep internet and the floor internet. We aren't going to speak about the dark net in this Book. But inside the next class we are going to talk a bit extra in-depth approximately the dark net. So allows begin All right So perhaps you found a picture of an iceberg on Google somewhere that announces it definitely is.

Well you have have been given four percentage of the internet. That is the floor internet. And then you definately truely definately have the deep net that is like 90 four ninety six percentage of all internet. Well this picture is a bit too large due to the reality technically the deep net is definitely every different a part of the Internet. It's like hiding on the net. If you are saying it this way, why is it this manner? Well it's miles pretty simple. The deep net includes all the pages that are not indexed thru a are seeking engine, as an example with the useful resource of Google. So what does this make? Well this consists of pages as an example YouTube Face book or each exceptional Web website on line that is listed through Google could be the ground net as I said or pages listed or the ground internet.

And then you clearly definately have all the pages that aren't indexed. So you have had been given tens of hundreds of thousands and billions and billions and billions of pages. And the ones pages are in the primary dull stuff. For example it is able to be your economic institution account. It can be your email. It might be pages may be essentially all pages which can be protected via password is probably as an instance a Website that is beneath production proper now. So as an instance a Web net web web page this is below production will now not be listed through way of a are in search of for engine. And if it is not listed with the useful resource of using a are searching for engine well because of this that the Web net net page isn't always at the internet yet. Well he's in all likelihood at the net however he is not indexed. So it's far part of the net and as quickly because it receives listed Well it is going to be a part of the floor internet. So you have the ones pages earlier than everything then you may have internet searches.

Those pages are basically antique pages of permit's anticipate in the course of pages we are capable to say what as an instance shall we say you have got were given got Yahoo or Google. Well they have got antique versions of them that exist however they may be truely no longer listed anymore. So the ones pages may be within the data. You can get admission to them via the usage of a few applications that exist to get proper of get admission to to older pages. But yet again they'll be just not listed anymore. And you have online content fabric. So it certainly is quite simple. It's all the content material cloth cloth on the net that has no one way links. So let's say as an example Web websites consisting of Quora wherein YouTube does Web net websites have plenty of content fabric cloth and that they've plenty of one-manner hyperlinks.

Basically it lets in them with you however it in reality is now not the priority of our Book one-way hyperlinks. Give your Web internet website online visibility as an instance you do. You located a hyperlink for your Web internet website on-line, let's consider on YouTube and people will see this hyperlink as a comments link. So Websites which have surely no lower back-hyperlinks and they definitely exist. They have perhaps some content cloth but they have no decrease lower back lanes. They have nothing. Well the ones Web web websites are very in all likelihood no longer listed via serps. And if they may be not, they will be part of the deep internet and as I stated earlier than properly the bulk of the Internet is in reality on the deep net so what sort of content are you able to discover it can be determined on the deep net. Once over again we aren't gonna.

We're now not speaking about the darkish internet. So the deep net as I said it could be any shape of content material viable and this is possible and that you do not want to truly properly you can for that the dark net you could need the specific software program application and the entirety earlier than the deep internet. Well it surely relies upon. Um nicely as I mentioned in advance than uh the shape of content material fabric material that you can discover at the deep internet can be specially silly stuff. For instance e-mails which can be economic organization account statements and those speeches are commonly included through passwords and topics that you can not actually virtually hook up with, so permit's answer a few questions on the deep net. So the number one one is that the deep internet is illegal.

Well no longer clearly. It's in itself not unlawful. If we talk about the dark internet or deep web or the net, both of them aren't unlawful in themselves because you can have a number of very very thrilling records right there that you could in fact find out on the floor internet. But all once more it's miles primarily based upon what you want to do with it. Some sports activities sports proper there can be taken into consideration as unlawful but all over again the idea of the climate internet is not unlawful however all what. How risky. Well how dangerous is the deep internet. Once over again it is greater dull than volatile as it's only a huge quantity of pages. So allow's keep in mind don't forget.

Well the everyday net is the ground internet is like permit's take into account a small library whilst the deep internet we could say it's miles a massive big large big library. So it's far commonly as I said boring stuff. So it isn't unstable however another time it is like if you're touring a Website on the floor internet that has let's assume viruses viruses computer virus malware as well you're as exposed as if you go to a Website that has viruses or my wares on the Deep Web. But we are not going to speak virtually in-depth about all of the risks of the deep net due to the reality yes you have some risks at the deep web which you're no longer constantly have on the surface net due to the fact nicely the pleasant of pages and the content fabric are specific.

Well some of you have got a few one in every of a kind shape of content cloth cloth at the deep net from time to time. But over again the most part of the deep internet is similar to the floor net. It's in fact dull stuff. It's the deep net. Same because the dark internet. Once yet again you apprehend it's miles each of them are brilliant. We are not gonna talk about it on this beauty, we're going to speak about it in our next class. And the final question that we are gonna suck is uh can Google are trying to find the deep internet. Well because it's not indexed on the deep internet considering that it's far not. Well for the reason that pages which might be on the deep net aren't indexed via Google. Well not truely due to the fact Google is not.

It's not taking this net page and people pages in interest because of the reality the bulk of them are properly personal or they do not need Google to index them and it's understandable for example an e mail issuer. Well they'll no longer want Google to index for instance pages which is probably alleged to be private. So the ones speeches are private and best strictly reserved to the person who owns an electronic mail. For instance once more that is part of the deep internet. Ok one final question. What form of stuff you may purchase at the deep lower darkish net. You can genuinely buy some thing. That's the detail, it is like a large marketplace.

Well you have got got were given pretty some marketplaces on the floor net. You have marketplaces which embody Amazon wherein there are various numerous things at the Deep Web. Well you have were given distinctive marketplaces you have got were given ordinary like in places like Amazon but you have distinct types of marketplaces in that you are gonna talk about it in our next education. But all yet again all of this is truly only for academic features. And in reality so that you guys understand that this exists. So this is it for this court docket guys. And see you all in our subsequent magnificence.

Difference Between The Deep Web And The Dark Web

All proper. So over again hiya. Yes and welcome once more to our Quest about the whole introduction to the Deep Web. Well deep reduce lower back darkish net face glass. We are gonna see the difference amongst every of these subjects. So what is the difference most of the deep and the dark net? Uh you are gonna see us. Um the distinction well there may be a distinction among the both of them as we said.

As I noted earlier than our remaining glass on the deep net is actually a place wherein for instance there may be no indication from search engines what this indicates. This approach that each one heights and seashores can be considered as pages from the Deep Web. It's let's assume one layer of protection but the Dark Web is a piece one of a kind. We're going to appearance why. So the Dark Web the way you may see it to start with it is nicely imagined the deep net. Right now the deep net has some unique layer of protection. So first of all to get entry to the Dark Web or the so-referred to as dark element of the Internet.

You want a completely unique software program thru manner of special software software we are talking about software program software application like Tor for example that's the most use software program software in the software program inside the most famous software program program to get proper of get admission to to to to the spirit of the Internet. Once again it's miles one and you've got had been given masses of various software program software that you may use no longer that an awful lot however you have got got some different software program application software program that can do the task like Tor on this elegance. We are in reality going to recognition extra at the excursion. We're going to artwork with it. We're going to artwork with every other software program however we're capable of mainly be working with our so you will want special software software program to get right of access to those pages due to the fact those speeches aren't dot.Com pages. They are dot onion pages. As you could see within the powerpoint.

Why is it like this? Well it is straightforward, you really cannot get right of access to them on a everyday are looking for for engine. For instance if you take an internet web page from the dark internet you could join the speech to a search engine. If you're taking an internet web web page from the Deep Web for example you want to get proper of access to a person else's Netflix account. Well in case you anticipate which you are all of the person's Netflix account you may write it down and you may be at the number one internet web page of Netflix due to the fact you want to enter the person name and the password to get get right of entry to to the beach of the person speech on Netflix can be part of the Deep Web. But subsequent Netflix as an entire like itself may be a part of this company. What is the darkish net? As I said it is actually part of the Internet that is hidden.

Once all another time it's miles not as human beings say a place wherein some of lousy stuff takes area. Yes you have got this part of the Internet. But another time it may be used very legally, very legally and for an super motive. So the 1/3 thing is that most seashores are very tough to locate. Why? Because you do now not simply have engines like google like google and yahoo as you have on the ordinary net. The way you could are looking for at the deep net is. Well on the darkish net is that you can really want to have the correct you are out of the region that you need to visit. It's very very important. And you can genuinely mess around and click on everywhere because of the fact properly it is not an extraordinary factor to do.

Used in particular on the dark internet it is a higher issue to do. So do you. You go to the Web. Well you want to have a selected you in any other case you may definitely click on on some random seashores or actually do your research as you do on the normal net. Because as I said the majority of things which may be accessible aren't boring and no longer very excellent. But don't forget the topics people say. But you have got were given some part of the Internet that is a bit more darkish.

Let's see it this manner. What form of information you could discover there. Other cryptic files you could have more cryptic documents all over again you want to have get right of entry to to the Web internet website on line you need to find out the places in which those files are. Sometimes it is able to be clearly in truth proper. Important documents you have were given prison nameless marketplaces. So fine you have got had been given marketplaces no longer as massive. Well we're smaller than Amazon but the ones styles of market locations you may find out surely something on the deep internet you can discover parents which might be selling

TV folks that are promoting mobile phones folks that are promoting all of criminal stuff. And yet again I do no longer understand why they promote it there, possibly it actually works for them. Well I do not determine it is their agency. So yeah you've got were given have been given a legal nameless market.

And the distinction all all over again some other difference a few of the well the floor internet reduce and a part of the deep internet and the Dark Web is that properly at the dark net the whole lot is anonymous. So it is not on the factor however the good deal at the Deep Web. Well at the Dark Web is permit's count on Anonymous because of the reality the software program application application that you have been gonna use this is stored properly it allows you live anonymous. Yes it's far now not 100 percentage anonymity because of the reality fine you could be tracked and these types of gadgets. But over again the motive of the deep net is virtually nameless. Well taking location the Internet anonymously we're going to talk approximately it in our future commands approximately the idea of anonymity. OK so genuinely for you to understand. Yes. You have a leak whole criminal nameless market.

Also what you've got were given is also a whole illegal. Huge unlawful community. So once I say large prison network you have got got were given sincerely the whole lot you can have it hold close from tablets guns to let's recollect for all the credit score score rating gambling playing cards. Well a whole lot of horrible subjects can be there due to the fact yet again the ones. This a part of the Internet there. There isn't any regulation there. Everyone is nameless. So human beings do quite a high-quality deal what they want. This is why there is lots of thriller around the deep properly of the darkish internet there. Well greater darkish net. There is a massive thriller spherical this trouble due to the reality people anticipate it is simply evil and outstanding that but humans will be there. But another time no it's far a small part of the Internet. It's now not just like the entire deep net like the deep net.

Let's say it's far ninety four percentage of the Internet however the darkish internet and the horrible subjects of the darkish net are best a tiny percentage of it now not as human beings say. But yeah you may have very very terrible stuff. For example, an unlawful network can have truely horrible stuff and may have racist content material cloth. You must have some very horrible Websites. So yeah. So it is it for this reason. Right now the distinction among the deep net and the darkish internet. Um uh besides that I do now not have anything to mention for this glass.

So see in our subsequent glass wherein we're gonna move a chunk bit more inside the deep internet and spot how this can be, how this will be used for criminal functions and what exactly you may do. Well what criminal topics you can find out on the Deep Web that would help you on your normal lifestyles. So see you in our subsequent clicks. Guys thanks for studying.

Chapter 2: Legal And Good Purposes

We'll see you. Hello men and welcome to three extraordinary of our education about the whole creation to the Web. So on this magnificence we are going to talk about how the deep darkish internet can be used for felony capabilities. Yes. Since I cited in our final elegance the deep internet cut lower back dark with the deep reduce Dark Web is not simplest horrible subjects. The deep deep net itself.

Well it can be severa exciting things truely well now not indexed via Google and the dark net. Well there too you have got got loads of thrilling matters. But all another time you have got got quite some horrific subjects. So let's examine how we are capable of do this. Two styles of net can be used for prison features. All right. So the first element is it offers humans anonymously. So the precept goal of the excursion and all this concept of deep reduce darkish net and the whole lot nicely the whole lot inside the again of that is genuinely becoming anonymous. So humans want on the way to have some privateness even as they're surfing. And with the ones browsers as an example if you skip on Chrome or Internet Explorer or some difficulty the whole lot you try this the extra the extra small movement you do well you are being tracked through one-of-a-kind cookies and in an effort to what is taking vicinity is which you are constantly nicely be big theorems the corporations in this international are continuously getting statistics from you.

So this manner they come up with lots of advertisements and you purchase their merchandise and all this form of stuff however with the deep internet it is sincerely one among a kind. You are a hundred percentage anonymous. No one tracks you. You are shall we say unfastened to do some component you want so it truely is an remarkable element. It offers anonymity. Yeah Of direction masses of humans why human beings are so terrified of the deep internet or the Dark Web is because nicely they are used to solid locations for example. Well Google it is what Google or every different browser net or Mozilla. Well they're secure places and Websites that you can locate there are safe Websites and generally you do not count on to look like stuff that you could see at the net due to the fact people will not placed up this form of stuff because of the truth they and they may be able to get tracked but at the deep cut back their internet it's far unique for the purpose that everybody is anonymous.

Well human beings put up any form of stuff. This is why you may have get right of entry to to. Well you can have get right of entry to to that stuff but you could moreover have get proper of entry to to fantastic stuff. And what I'm speakme about is excellent stuff. We're gonna speak approximately it. And our final trouble. Also it can be real in any kind of facts. The 2d element it offers access to is well censored content material material to humans from a few global places. What does it suggest? Well what I'm trying to mention via this is shall we embrace for example you've got have been given international locations that have a few Web net internet sites which is probably censored which you cannot get proper of access to as an example in some international locations in the worldwide you cannot get entry to Facebook wherein you can get right of entry to YouTube. So people who are dwelling inside the ones worldwide places are dwelling nicely or experiencing massive censorship that prevents you from gaining access to a few internet net websites. So what's taking area is that they do not see the Internet as we see it in distinct nations, for instance in Canada. Yes.

In The United States we have got get right of entry to to content fabric fabric that humans in China will not necessarily have access to because the Internet is a manage. So the the deep internet nicely the deep net the browser that that properly that controls that gives you get entry to that a deep net will provide you with get right of access to to all the records and you will be capable of nicely get get right of entry to to pretty a few content material cloth that is censored with the aid of manner of governments and you will be capable of study this content cloth this is an tremendous issue. And once more it brings us to the aspect that a number of statistics is available on the deep net given that it is around ninety 5 properly from 140 96 percentage of the complete Internet. You have first-rate brilliant content material right there. And for the censors. For the censorship component even this for instance people were residing in a positive u . S .. Well they may be capable of see the area differently.

Let's say the government attempts to no longer oppress you but the authorities attempts. Usually in maximum people's well the government controls media the government controls the entirety and through gaining access to the deep net gadget human beings can be able to see the arena like they'll see the arena in a exclusive way. They will now not satisfactory see what their government or media is telling them to look. Let's anticipate it has what it brings us to our last element this is it has hundreds of big facts. So what I'm attempting to say through big facts properly first of all you have got all the ones Websites that humans cannot get right of access to can be capable of get proper of entry to by using the usage of going to the Web. And you've got were given got huge libraries, you have got books, a wide variety Of guides you've got were given.

Well whilst you get the deep internet you've got one element that is the tour library I'm now not mistaken. And properly you may be able to have get admission to to a large database Of publications that you may look at. So it is lots of facts. Also because of the reality that it's miles like the most important a part of the Internet. Well by means of having access to the deep net lessen darkish net you'll be able to get get right of entry to to pretty some lovely data from taken into consideration one in all a type worldwide places. Well specific people best really do the idea of having access to what other human beings say you're the point of view of some people on a few topics or some conditions. Well it's simply a beautiful component due to the fact over again humans are nameless and people do now not hold their sight of seeing the real things like what they honestly assume.

And this manner you may once more see really lovely matters but you may additionally see very terrible subjects. So that is why it is very important to apprehend wherein precisely you may look to discover your statistics because of the reality as I stated you do now not have censorship on the Deep Web. Well the huge dark net because of the reality absolutely everyone is anonymous so no censorship technique people can say and do some thing they want without usually getting tracked. Even if a few humans have emerge as tracked, that is the concept of the deep dark internet. It's in reality to present you the Mahdi and provide you with the electricity to say topics or do subjects without being censored. So it may be a high-quality thing. A very incredible issue but it is able to moreover be a bad element.

So that is why you need to clean out in which you attempted to discover your information. So the ones are the things that you can do spirit of factors. There are ways as a way to use the deep web legally, as I stated, to discover information. Once again it is able to be scientific statistics it could be facts about some of distinct stuff about history about nations. Personally I in truth love that excursion library because of the truth you have books approximately certainly the whole thing. So yes that is in my opinion why I use the deep internet very very often because of the reality as I stated you've got pretty some lovely facts and you clearly need to understand in which to discover this statistics and a way to filter all this records because of the truth positive every now and then you could discover very awful subjects.

But yet again in case you filter out right, if you smooth out the facts right away the whole thing's tremendous. You're proper. So those are the three things that for me I recollect may be considered as well you may use the net for crook purposes anonymously to provide get entry to to censored content material fabric cloth for it's for nations that stay in censorship. So those human beings may want to have get proper of get admission to to to the whole Internet as an entire and could deliver get right of access to to an entire lot of information that isn't always necessarily at the ground net however that may be very gift on the deep cut down dark net. So this is it because of this guys. And see you in our subsequent magnificence wherein we communicate about myths about the dark slacks, the deep lower.

Deep Web / Dark Web Myths

It's quality been a few years. Welcome once more. That is Of path our Book, the complete advent to the dark internet. So Of course we're going to talk approximately the crucial myths and legends which are all through the Deep Web. I'll provide you with a few examples and a disBook that you may see at the issue and I'll supply a few unique examples as nicely. So permit's begin, the primary myth is that the Dark Web is massive.

Once another time I'm no longer speakme approximately the deep internet. I'm speaking about the darkish internet itself. So the deep internet is as I stated something spherical 90 four detail 80 six percent of the complete Internet. The Dark Web, the darkish net himself. Well it's no longer that huge in comparison to all this. It's exceptional a few factor approximately some aspect round 500000 Web net web sites. So yet again it is very very small in assessment to the deep internet as an entire. So all over again even as we communicate approximately the Dark Web it is all the subjects which can be hiding beside the network and all of the topics which you want.

Well every other you need precise software software application is to get get right of entry to to so all over again it is no longer that massive. If we look at it to the floor internet or if we observe it to the whole deep net as an entire. So over again it really is from property that you could find out on. Well in case you do your studies the belongings mentioned that it is now not that big proper. So the identical detail. The Dark Web is unlawful. So and so now you recognize that the dark web has some very horrible matters in them. But another time it's miles now not illegal in itself. Getting access to the internet is like getting at the darkish internet and stealing this software program program and getting on. I do now not realize that any sort of Website on a dark web is not unlawful. It's what you are doing after that.

For example you in fact get at the darkish internet and mess around and visit. Well it isn't unlawful however if you try to shop for it shall we say weapons. Well if this is wherein it turns into prison. So the dust as I said dark internet has itself isn't illegal. It's even jail to move on it due to the truth. Well as I said, my tremendous commands need to have very exciting facts there. And well you could use it for a whole lot of instructional stuff. Well academic competencies as I said you have got got got plenty Of publications you have got loads of factors of view on an entire lot of various matters. It's truely an superb region. If you recognize the way to apply it due to the reality in case you do not know the manner to apply it nicely it could be very awful.

Third aspect is that the Dark Web is completely anonymous. So sure with the resource of the usage of the software program application is related to having access to the deep cut back dark net you're alleged to be anonymous but another time it's miles based upon what you are doing on-line. So there are plenty of human beings which might be attracted via way of the dark internet and no longer all, not all of these human beings are there for the right reasons. So you can have hackers. That's a huge area for hackers and sure you may get attacked. Yes. Depends what you're doing. There is a few form of stuff that if you do properly you can get in hassle with regulation enforcement, as an instance getting attacked with the useful aid of different human beings on the dark internet due to the reality over again if you are there simply check matters and simply to have a look at books and to get information.

Well it isn't a large deal. But all all over again if you change it it is it is after which it's nameless till a person wants to music you. For example if the law wants to song you due to the fact you have carried out some issue illegal. Well then it isn't going to be an sincere guy nameless anymore because of the fact they've got the price variety to find you so yeah it's far anonymous but handiest a effective component. It's not completely nameless. No, there may be not something on the Internet that is a hundred percentage anonymous.

Dark Web is exceptional used for crimes. So till now I assume my element of view approximately the dark internet in case you recognize a manner to use it the right way. Well it is not a awful issue. Once all all over again as I said you could find adorable records. You can discover books that you can't usually locate in libraries or books which is probably very very tough to find on Amazon. You can find out PDX of those books. You can locate certainly brilliant information there similarly to the boards on an entire lot of special subjects which may be political and monetary about governments. There is not any censorship. So once more it is now not best for crimes it's miles for human beings like you and me who simply want to have a look at new subjects.

Yes you've got were given got criminals at the darkish internet however you have got got got even more criminals at the ground internet. Why is it like this? Because on the ground net there are lots more humans. So if we talk approximately cyber unlawful interest agree with me the hackers is probably more observed on this at the floor internet at the deep net because of the fact at the Dark Web net web site due to the reality the darkish net does not have a whole lot of internet site site visitors in contrast to the ground net it truely is why hackers have a tendency to assault more ground internet internet internet web sites than darkish internet internet net web sites as it's greater hard to assault those Websites which may be on the dark internet because of the reality all over again there isn't always that a whole lot website web page site visitors and every person's using the Twitter community and it is way extra greater hard to obtain the ones people than to reach human beings on floor net wherein the records is manner extra to be had.

If we communicate about crimes like now not cyber crimes or real crimes. Well it is the appropriate equal trouble. Criminal criminal sports seem greater on the floor net than they appear inside the dark at the deep cut lower again dark internet. Because all once more there are way more people at the ground internet than at the dark deep reduce down darkish net. So yeah it honestly is in which the crimes and over again as I counseled you earlier than the deep dark internet may be used for a whole lot of acceptable stuff for an entire lot of top abilities. So for me it isn't always an area for crimes and for some of individuals who recognize how to use them. It's no longer an area of crimes. It can be once more a superb place for facts.

It's very hard surely there. But now it is very easy to get get entry to to the Dark Web. What's difficult? It's navigating the Dark Web proper with submitting. Well, finding Web web web sites or hyperlinks to precise Web websites it is very clean. But over again you do not need to get at the Web web sites in that you have viruses and your laptop gets attacked or shut down because of the fact over again it's far no longer a criminal offense. It's no longer an area at the programs. But once again in case you click on on the wrong link properly you can get on the Web internet sites that you don't actually need to get on. This is why it isn't difficult to get admission to it however the use of it the proper manner. Well it's far a bit extra tough. It's gonna take you a bit of time to find out locations in which you could find out the topics which you're looking for. For instance, you need to discover records.

I do now not realize about permit's bear in mind political regime of a superb america of the us of the us. Well, do you have got got specialised Websites there? Let's say you want to discover a e-book about I do now not recognise the sea. Well you have got had been given places on the deep internet in which you may find out this e-book in which books associated with your trouble. You have locations in which you have an entire lot of statistics. So all over again it isn't always tough to get right of entry to the dark net however it is more difficult to recognise the manner to navigate the proper manner. True. Well the proper manner through the best darkish internet. So you do now not really need to be an expert to get get right of entry to to this Web internet page. So it's that for all the myths which is probably across the dark internet.

And then another time it is very critical for you guys to not take into account all of the creepy pasta that you may see on YouTube or some different platform about the deep darkish internet because of the fact loads of the ones people are saying nicely we are seeing loads of things about deep the ones darkish internet is that they may be by no means been there or they've been there for like 30 seconds every and that they absolutely depart. Leap because of the fact all another time it's far if you recognise the way to use it the proper way. Well it's no longer. It's more stupid. It's much like the ordinary internet however simply Web web sites are much less and plenty much less permit's expect flash you are a whole lot much less appealing. That's the most effective aspect. But besides that if you apprehend the way to use it.

Well it's miles a in reality tremendous area to investigate new subjects. And well most of all research new topics and observe new books. So permit's do that Book men. And I want you loved it. See you in our next elegance wherein we are going to speak approximately the software program application software that we are able to use to get get entry to to the deep net.

Introduction to TOR

All right. Once all over again Hello men. Welcome once more. Of course the deep net is the whole creation so into this beauty we are going to have an advent to tour. Um why tour. It Is quite simple to get right of entry to the deep internet. I. Well the dark internet. I've knowledgeable you that you'll want to use software program software , so the software software that we're gonna use these days is the maximum well-known software that exists to be had available on the market. This software application software program is stored. So that is why we are gonna speak approximately it. So to start with what maintain. Tourism. No. Nope source task.

This means anyone can make contributions to the task. The fundamental aim of this assignment is really to sell anonymous navigation and then flip it this way humans may additionally additionally have get entry to to data that they need while not having to show to every body who they may be. So it simply permits with how the browser works. It's pretty easy. The excursion browser works on your network. So at the same time as you be aware an onion what do you notice? Usually an onion has hundreds of layers. So this is exactly the same thing. When we communicate about Tor you've got got Tor and allow's count on as an example you use some other internet browser due to the reality your save is a web browser first. So let's consider you operate for example Google Chrome whilst you use Google Chrome and as an instance you want to get get right of access to to Google.

What's going to appear is that Google Chrome is going to deliver a request to Google server and the request will come lower lower back and come up with get right of entry to to Google with the two browsers. The request is gonna pass so that you will deliver a request from your computer but the request will skip as an instance to the server wherein a pc is remote places after which a few one of a kind america of america and then a few other u . S . A . After which in the end move decrease returned to Google. So what actually happened is that tourists despatched it. Well in case you send a request by means of a number of computers this way does find the person who obtained that request. It's Google that obtained the request and the request will do. Once all all over again the entire journey lower lower back to you. So why is that this completed? It's pretty simple it's far as it's a super way for now not getting song at least on a critical diploma due to the truth well the request passes to a number of laptop reduce servers and this manner humans properly people are making studies at the Internet and manner more extra tough to track than those who use for instance Chrome or Mozilla in which net or only a simple VPN.

So if you ask a few questions that people ask, for example to cover my IP deal with certain or thru passing to unique pc systems, decrease servers will disguise your IP cope with. And it definitely is one of the important desires of this software program application is absolutely to help you live anonymous on line and by means of the use of hiding your IP deal with is one way a good way to stay nameless online due to the reality no one will recognize in which you're and from in which you're browsing. Um positive Tor is a browser for people we're asking. So it's far the equal aspect as Mozilla Firefox or Chrome or something. The exceptional distinction among this internet browser and the possibility ones is that this one is focused on surely staying nameless. Is it loose?

Yes, garage is completely unfastened. As I said, an open source mission. So it is sincerely loose to get get admission to to it. And no it is not unlawful to apply the shop for. We're asking the deep internet over again that it isn't unlawful so the software this is spherical having access to this nicely to the net web page online of the internet is likewise now not illegal. Once all another time. Well why the browser changed into born in truth sincerely one of them after which one of the vital reasons why that net browser have come to be born is in reality to assist solution in want of anonymity at the same time as browsing online. So quite a few human beings are internet surfing on line and want to live nameless. And it is surely surely tough to be nameless , particularly on-line.

Because at the same time as you're browsing properly despite Tor if someone wants to music you down there are methods to song it down. So with exceptional browsers, it's miles quite easy to track a person's down. This is why the net browser became born. Because it's miles virtually to help people live nameless on line while now not having every person understand what they do. Exactly. So it certainly is why the browser existed. And as I stated the principle goal of the excursion browser. Well it's miles pretty easy. It's getting people supporting human beings stay anonymous, mainly online. So it truely is the precept goal of the door buzzer. The browser itself. Well the excursion project started in 1980 1990 and it become honestly advanced round 2006 if I'm no longer mistaken the primary model got here out in 2002.

Well the initial launch but. Yeah. So the task itself has some difficulty in reality stunning within the again of it. Because as I said it's far certainly to help people stay nameless due to the truth as we understand in modern-day society absolutely every body's monitoring sincerely all of us. So our undertaking is clearly beneficial for this. And but besides that the tor browser as I stated is certainly to assist anonymity. So proper now we've got the identical shape of concept about what maintain to art work in if it is prison or illegal. And so as I said it's miles no longer some thing everyday, it isn't some element that scares me that a web browser is honestly another browser much like the unique ones.

Well as a few different browser the best difference between this browser and others is surely that it's far centered on staying anonymous and helping the individual that uses it live completely nameless on line. So it truly is the furthest Book gays can see in our subsequent tourist wherein we're going to talk approximately the protection within the again of tor.

Chapter 3: Tor Security

Hello men. Welcome lower back to a few other of our Books approximately the whole introduction to the deep internet. Since this magnificence we're going to talk approximately toy protection and a manner to use Tor very well. So permit's start. So what's idea of stands for The Onion Router. It comes nicely that is a internet browser as we use as we stated it. Now the closing test. And offers the bot the opportunity to people who use it to stay nameless online and to get get admission to to the Deep Web.

But is it safe to use? Well yes not typically it's regular to use. But you need to take some safety measures. Once again um Tor isn't always some aspect that prevents you from getting viruses or mail malware or some thing to position on what I'm speaking approximately nicely. I'm talking from an nameless factor of view. So if you want to be nameless or if you want to be permit's expect under a piece whilst you are taking place the Internet sure it does a pretty hundreds precise challenge it is not a one hundred percent because of the reality no net browser can guarantee you a hundred% anonymity. But once more it does a pretty top notch process. So for death, so for that encryption.

So the way to guard your anonymity. As we noted very last beauty. Well it makes use of an onion the manner it truely works. It works like an onion. So it's going to begin at your region to begin and then allow's bear in mind for example you want to go to Google. Well it's going to bypass to the right shall we embrace pc systems. It's like an onion while you're commencing layers from an onion at the identical time as there are masses of layers. It's definitely tough to get. First the let's say layer or first man or woman who as an example preferred to get get right of entry to to Google which in this example is you due to the fact you are related to Tor and you need to get right of entry to a certain Web internet web page. So let's see this. So for the statistics encryption element Well it's very very very it's miles it's far very it's very but it's far pretty a notable deal pinnacle. And nice. So it actually works quite nicely.

Once again Tor would possibly no longer save you the whole lot from end to stop. So. Well from one bird to the alternative allow's recollect from your net website on-line to the Web internet internet website online you need to connect with and you can should be cautious due to the reality all another time there are people who are able to music you if you carry out a little terrible matters at the

Deep Web. Um, so how do they're announcing the first factor? Well, equal issue. I recommend you need to be careful with some plugins so even as you are related to the tower browser you do not want to down load a number of plugins that you could discover at the Internet or on YouTube films. Why? Because Tor is already finished. So whilst you download this tool it's miles already completed. And in case you down load plugins properly it can intrude with the interest that is already carried out by means of way of the browser it itself works quite well. So you do no longer need to upload pretty a few plugins to make it paintings better. It can.

It has greater possibilities to area you at hazard than to help you with some factor. Because none of those plugins is assured to be robust. That's the element with all of the plugins that you download to do it. Next issue is when you're navigating on the deep net and also you want to be a hundred percent sure that you're nameless. I in my view advocate you close all the apps that are linked to the Internet in your computer. So it's my belief. I in my opinion do it. Um why. It's because of the reality nicely humans say that they will be what human beings say. There are ways that humans can track you if you go away those apps open so it is best a tip that I for my part use.

I do no longer depart the apps open on my computer as soon as I'm navigating on let's assume the deep net or after I'm connecting to the TOR browser genuinely so I'm no longer a safety tip that I can offer you with. If you ever navigate on that internet browser. Um. Next issue on the identical time as you will be associated with a browser the Tor window will. It's gonna. Well it's miles gonna be a tremendous duration so do now not maximize the size of this window. Why? Because thru maximizing the size of your window you are setting yourself at risk that you can get tracked or your anonymous city will lower is while you can. Well in which we may be at the issue wherein we've have been given a download Tor you could see this morning on the equal time as that internet browser can be downloaded.

You will see the caution that you aren't constant at the same time as your window is maximized or Max um and so this is why. That's a few different protection tip that I can provide you with. Um do no longer maximize your excursion browser window leave it as it is. You will see the dimensions of it is not that small so you will be able to navigate correctly out of your laptop on your browser on the equal time as no longer having to maximise it. Um next problem: Never ever ever download files on the deep net or cut returned. I endorse darkish internet Why. You realize what's within the documents. You do not recognise what number of viruses you could download, especially in case you circulate on a streaming internet website and also you want to download films on the darkish net.

Well it is a completely horrible idea to down load films there because of the fact there may be more opportunities that you down load a few virus or some malware as a manner to connect with your laptop and steal all your information that you down load best due to the fact all you all the ones Websites inside the ground net are complete of classified ads and with viruses. So at the deep internet it is even worse. You have even more chances to get an epidemic thru your computer so by no means download no facts. Never. If you get hold of an electronic mail on your nicely to your tour under a door for your Deep Web message carrier do no longer. Well you can open it but in case you acquire for example attachments to ask you to download some aspect.

Well do no longer down load it at the ground net. When you get keep of an email of some thing as a person who you do not understand asking you to down load a few aspect. Well the superb element to do is not don't forget it. Well it's far the identical issue on the Deep Web. Well do not do it because of the fact you can see you're gonna get your self in masses of trouble specifically in case you do now not have the purpose of it. And antivirus or if you're not the usage of a virtual device. And in this Book we are not speakme about connecting your finger properly within the course of our digital tool but I speak approximately tables but it's miles a bit special. But yeah I.

Let me absolutely provide you with a quick. Well, virtual machines. If you join through toward the digital system it is a piece exquisite due to the truth you will be able to. Well in case you trap that you say the viruses will only affect the digital tool and could no longer have an effect in your pc. Once another time on this Book we're great talking approximately getting related to the deep internet, not the use of lots of stuff to be one hundred percentage nameless. Yeah. You said that. Don't mess around with the darknet at the same time as you are on the darknet. You need it. Well what I advise to you for my part is you could need to. You will want to have a list of Websites you need to visit.

Let's say as an example you want to visit this Web net website or this Web web net site. Well wherein you're going precisely in that you are. What precisely you need to go to. And in case you want to do some studies as well. Make them at the clear internet or the floor net or. Well you could cross on certain forums that let you with this. But do not start looking spherical and clicking on all links that you can discover on the deep internet because of the reality there are more possibilities that drift on an undesirable Web internet site than some thing else. So another time be very very safe in this course. So you could need to have a smooth Website in that you want to move if no longer what Matrix researches on the floor net for the hyperlinks and then only after that may you change those hyperlinks at the deep internet. So permit me summarize another time that our encryption for the nicely for the safety it's quite an entire lot suitable.

So it genuinely works thoroughly the right manner. So be very cautious with the plugins that you can down load for the internet browser that doubled now not something from the the front while you visited the deep internet. Nothing, not anything, never closed our apps even as you have got been using Tor so bear in mind about this all up close to. Only the 2 hour window opens to maximize your door window. You're too moist to window because of the reality it is able to be tracked if it is maximized do not open and download files or do not down load no document.

No begin downloading stuff due to the fact there may be viruses, malware or any sort of belongings you are now not on the floor whether or now not you're at the Dark Web so you want to be very careful it might no longer start to mess around all over again because of the fact you could get on Web web web sites that you do not need to get into. So this is where this Book is going. And then I see you all in our subsequent elegance wherein we are gonna communicate about the idea of anonymity.

Chapter 4: The Concept Of Anonymity

All proper I'll see you yet again short. And welcome once more to a few different of our training Of direction about the deep internet. But in extraordinarily-current elegance we are going to speak approximately the concept of anonymity and excursion alternatives. So at the same time as we're speakme approximately the idea of me I really want to talk approximately why on anonymous why and this concept as an Oh. So permit's begin.

So the primary question is why human beings want to be anonymous online so there can be an entire lot of reasons for this. First of all, humans sincerely want to be anonymous. People virtually do now not want to be spied on or to experience that a person is asking them or some thing any kind of purpose related to this second purpose. For instance, humans are touring internet web sites that are censored in fine international places. So as I mentioned in my past training there are global places on this worldwide in that you cannot get get right of entry to to three Websites. For example, a few humans can cross on Facebook in some nations. So let's anticipate they use the phrase browser.

Well they'll be able to get admission to Facebook or every other internet website online for instance in some international locations. There are data web sites which might be banned. There are journalistic Websites which might be banned. Well through the use of the Internet browser you may be able to get right of entry to all the ones Websites virtually due to the reality you're. Well you'll be capable of get right of access to all those Websites so the ones are the essential reasons why people want to be anonymous. Also if people need to put in writing some aspect allow's expect anonymously. Well they're able to do it on the deep net due to the reality. And thru manner of using that net browser does.

If they do it at the floor net. Well if dwelling in a rustic wherein censorship is slightly heard. Well they may be in masses of hassle. So it simply is why the following element. Is it feasible to be nameless on-line? So despite the fact that we are nicely proper now we are talking approximately the concept of anonymity. But is it possible to in reality be nameless on line? So for my part I assume no longer because of the truth the positive story gives you a nice anonymity and the entirety. And in case you're using a VPN or some other problem that hides in which you are. Yes you're nameless till a excessive fine element but in case you are doing topics in case you're doing illegal things or any associated stuff to this. Well there are quite a few opportunities that police or Internet Internet police will want to track you.

For example if you're a hacker and you are hiking or something. Well there may be quite some chances that the police on the whole internet will need to song you and that they have got an entire lot of budget. The difference among you and them is that they've plenty of budget. This way they have got a whole lot of resources and they're able to even in case you assume you're anonymous properly they're not in reality that nameless. So positive there's a sure anonymity at the internet but till a sure thing all another time if someone absolutely wants to track you and has the belongings properly it's feasible to song human beings even folks who are the usage of the internet browser with. I do not apprehend how many layers of safety there are. So yes it's miles viable to get on the mass online however you in no way may be a hundred percent nameless. That's the handiest detail.

Next question using it once more in area of tor browser. So essentially the distinction among a VPN and the excursion browser is that a VPN will normally be faster and uh due to the fact they've got their very very own servers. The primary cause for that is due to the truth the chickens have their very very own servers to type and Tor is a bit more plenty slower because of the fact change is based upon on folks that. Well people who are volunteers too properly get the release of allow's expect you will Google. Well you want human beings to really need to visit Google. So they depend on volunteers. But the plans have servers

VPN organizations have servers anywhere within the global. And it really is why they're loads a whole lot faster.

The simplest trouble is that Tor well tor proves that it without a doubt works and proves that it is some component clearly very dependable and furthermore tor it's far free. So you do now not need to pay for it. So in my opinion I like Tor and simply whilst you're taking region the deep internet only for example allow's don't forget proper read books or observe a few type of data which is probably censored in a few nations. Well with tor you will be good enough. You don't want any fancy VPN or each other shape of stuff. And do not attempt to use each VPN to tour or. Because typically as I referred to my last elegance um talk comes with already built in capabilities and the tour mentions it uh noted that while you set up it. That it's far no longer a superb concept to just what even as you installation Tor to function unique third birthday party cloth to to 1/3 celebration plugins or add ons.

It's no longer an super concept due to the fact Tor already has his topics and until now it simply works pretty well. So proper now is there any possibility to Tor or no longer really. Once once more it really is a first rate query and which you have hundreds of options. But for me for my part I recognize 3 of them which is probably which might be very that works well the primary one is that that works very well it is also a peer to appearance platform. It permits. Well it has the equal motive as that in which uh it tries to nicely it tries to now not promote censorship. It tries to decrease censorship inside the global and the international places who are censoring a few as an example records, a few records or some topics. This platform, permit's name it like this, is clearly closer to it. So if you're searching out a few factor permit's expect Tor loose Net there can be like this.

The 2d nicely let's talk about the attention to be the second one to be is invisible Internet challenge. Once another time that is every different issue that is a few thing like Tor. It's an anonymous community layer. Once another time it has 55000 pc structures in the works a chunk the equal manner as Tor so it's miles launch. So as an example you bypass on Google and it honestly works precisely the same way. So there are releases so you can pass through masses of computer systems. So this way no individual may be able to bypass again to you. Well it's miles gonna be tough to get all over again to you. And the closing little detail. Well the final opportunity to tour. Well it nevertheless is but it is not genuinely a no turn possibility to excursion because it works with Tor. And we are gonna. Well we are gonna download it.

Also we are gonna paintings a chunk later in this splendor of lodges. It's quite simple, it's far a lifestyles device. And whilst his goal is also to maintain private Senate nominations anonymously, this device works immediately with Tor and Tories blanketed into Tales. So this manner not simplest nicely it is a primary gain to have it because what is taking region with tails is that you will have it on the US booking so you may be capable of get right of entry to the deep net everywhere. So shall we embrace as an instance you are going. I do now not recognise. In it, a vehicle is in a espresso preserve someplace and you have were given your laptop for a. I do now not recognize. You have a computer or someone's pc.

Well we will take you to have offers on it. Plug the laptop into the deep net, allow's consider observe a few records and we'll certainly unplug it and all the records will now not be stored anywhere. It's going to be. Well it is gonna be raised so Thales is in reality in fact an fantastic device to use. And that is why we are gonna discover the way it truly works. The best element about it is that the whole lot that you men are going to see on this Book on the same time as all of the cloth that you men are going to use in this Book includes tor paintings gives is actually unfastened and it genuinely works very well it's been examined now not best with the useful resource of using me but via way of quite a few human beings on this planet. And it is also all prison and it works nicely once he works very very very well.

And for individuals who want to live without a doubt anonymous or just navigate on the clean net anonymously. Well they will be gonna see art work. It's a fantastic manner to do it. It's. It costs no longer some thing. Yes. It's gonna be a chunk slower due to the fact you will should wait because of the reality as I defined towards the manner it works it's lots of folks who. Well even as you're making a request to a server it's far gonna in reality buy some of computers and this is one one of the motives why it is a chunk gradual. If you have this if you have money and also you want to pay for a VPN and you want to stay on the clean net properly you may do it. But in case your intention is to move on the Deep Web.

Well you'll see Tor and tails. They is probably your super pals. And in the next training or maybe after this Book. So I expect proper now you understand what is the idea of anonymity and why people want to be anonymous. So what are some alternatives? Yes you've got got some distinct options. But for me those 3 are quite an awful lot first rate. My preferred is tales that art work right now with doors. So positive it's miles a thrill. But it furthermore works with Tor so it isn't virtually a lavatory. It's like some issue that works with Tor. So it is it from this class men and they may be seeing our next magnificence wherein we are gonna pass on the sector and we are gonna start jogging with Tor and we will installation it.

Downloading TOR

All right so all yet again how little men. And welcome lower lower back to our 9th magnificence or Book about the deep internet, the complete introduction on the grounds that this beauty. We are going to download the tor browser on our pc and we are going to set it up so this way you will be able to use it for. Well as a personal nicely because the primary browser we're clearly as brother brothers at the deepwater. So till now you apprehend all the theories which you want to recognize so one can down load. Well that allows you to navigate at the deep net. So allow's start our adventure right now.

So the number one element you need to do is open up your normal are seeking for engine. If you have Chrome will open up chrome then quite simple you need at the same time as you are your are seeking for engine truly. Right. Don't communicate whilst it's completed. Well you may click on on on the primary link. It must seem like this. So it'll probably be achieved now and also you click on on on Download the browser whilst it is performed. Very genuinely, actually choose your system when you have a window. Download it for Windows Mac from Mac and X and Android. So you sincerely click on on on down load and it's miles going to be downloaded for your tool. When it is completed you need to have a component like this.

Well you need to have this step so one can seem that is begin tor browser. You definitely click on on it. It's going to load. You'll see it's far a chunk slower than special browsers however it's because you have were given some of. Well you have got were given at once on it that makes it greater strong. So we are able to see probable the number one time that you down load it. It's gonna take a piece extra time. All proper. So proper proper right here is our tor browser. So it ought to appearance some detail like this. So the number one element you need to do men. Well the number one problem we can see is how it's far. So as you may see it is similar to the Mozilla net browser. Well the look is a piece the equal. So let's see the primary problem I need to search for is what are the upload ons as you can see you've got got already upload ons on your browser.

The first one is H2. Yes, everywhere. So this way Oh the Web web sites that you will go to have to have a properly automated protection on them so they'll automatically generate a effective. Think then you may have every other thing that has no script. So that this wait will boom your safety while you navigate on the browser. Next issue we want to do is on the safety stage. So simply click on on there. Then we will pass at the safety diploma from here. We're going to speak about all of the setup of the excursion in case you want to set it up as your default browser.

Simply check proper right here. Once yet again as I stated it is a normal web browser as another. So you can use it as an everyday mission. It's a piece longer due to the truth nicely it's miles longer as it's extra permit's assume nameless but you may however visit Google or some other Web internet site on-line with it all right so proper here in case you want to exchange the arrival of the excursion if you want to trade the language is going to be proper right here. Here click on on the second because you continually need to decide in which you may save your documents. So allow's do not forget you need to store a record to your laptop after which underneath record at each other vicinity.

You can do it or what you may do is also shop documents too. And then you definately definately create every other report that you'll call Tor and all your documents which you downloaded from let's say the deep internet. We'll be safe there. But yet again do not down load documents from the Deep Web. But in case you do it then for the updates it is gonna be right right here. For instance if you need to download updates for your tour brother it's gonna be there. So you decide if you need to donate them robotically or if it's going to ask you to download them. Uh this little element here is home so it isn't always very important. It's simply to allow making a decision what's gonna be your homepage so you can add custom if you are sick. Or it could be the net web page about Tor which is this speech. Same detail for New that subsequent the search. So right here you may decide what is going to be your default are trying to find engine.

And in the end the protection page it simply is the maximum vital internet web page. First element you want to click on on on on is proper there. So all your cooking information ought to be closed even as the is closed so you will haven't any Cookie left on your browser. When the excursion browser might be closed, log within the passwords and in no way go away your passwords anywhere. So your passwords don't want to be stored so do not take a look at the ones subjects proper right here. Once again for statistics. Always use a non-public surfing mode. So check this element there after that for all of the permissions. Very easy. You click on on settings and also you test this field proper there so block your requests. This manner no person is probably able to supply your requests if allow's assume you're on Skype but your I do not know you are doing a little face time from the excursion browser nicely you can probably prompt it but at the same time as you are surfing the deep net virtually block all of everybody that might deliver you a notification to get get proper of access to on your area digital virtual digital camera make or notification.

Same issue proper here. So take a look at up all the matters right there to block net websites from automatically being sound pop up home domestic windows and the only if a internet site desires to installation upload ons. This is for example if the website wants to routinely installation some thing for your pc. Well you can block it there if you need to feature some exceptions whilst you can click on on there and also you upload the exception Web internet site. So that is as an example your streaming movies on line in which you circulate on the net net site wherein there are masses of dad united stateson the floor internet.

You can just check this field and you'll have no pop americaon your pc which the rating room then to your safety diploma you've got got degrees of protection stylish. This one is whilst you go to allow's anticipate the surf Web helps you to use your personal google or some element you're taking. This one right there. If you want to have a greater constant ID you can test a second one will do. It will disable all javascript anywhere but anywhere story on most effective no no HD s Web net web sites. So let's assume for instance it's far a dot in your Web net internet site that isn't always consistent. Well it's gonna be disabled.

The javascript may be disabled so you may not see pictures and different subjects. Same difficulty for a few symbols and well a few budget and audio video may be clicked on the click to place. For instance we've got were given a video. You will want to click on to play films may not start in case you click on on on them and the remaining one will. It's the most steady one. It's javascript is blocked anywhere so there can be no javascript you cannot have any attack in your computer. A lot of the icons and logos and photographs are disabled furthermore. And ultimately the equal component for audio video and the internet. G L Well it's far gonna be click on to play so in case you want to as an example watch a video you may must click on on on it. Then to check all of the ones tree bins there it's far to block you from unstable and deceptive content material material fabric. So you need to check in and the forces which may be tough.

Check the second. So it is gonna ask whenever for the primary and for the 5th that right proper proper right here you have got were given not a few thing to check it is a chunk. Well more superior abilties if as an instance you have a proxy. But for now we do not want to use we do not want to use this feature right proper right here so that you can see proper there. You have Doug Doug. Doug the skip this is uh the net browsers with the Twitter browsers search engine. It's a parade. Well it's miles anonymous, a are seeking for engine and that you moreover have a domain where you've got have been given a dot onion model of duck duck pass for now allow's have a have a look at what it looks like whilst you are making a are searching for on it. So let's consider we search for Duck Duck Go.

The Web version. So as you could see takes a piece greater time however you may see it gives you all easy that the outcomes so permit's click at the quit cease result go. So as you could see it's uh properly this is the clean internet version of Dr. Because it's far a dot com Web internet site however you surely have a uh model that is uh for the deep internet it without a doubt is a version that says Dot onion. So allow me display you wherein it appears a few component like this. So as you can see this Web net website online is uh properly it has loads of numbers and letters.

It's not a traditional Web internet website online on line and it's far a dot to your Web internet site on line. Once all over again it's miles in reality the D version of the satisfactory this is completed. Well this is at the Deep Web. It's a dot in your Web net internet site on-line. But besides that it's the identical dumb that is happening leads as I said a more nameless model and say it this manner. So it clearly is it for the rankings guys in you all. In our subsequent beauty wherein we are gonna go to some Web websites at the Web.

Chapter 5: List Of Safe Websites To Visit

And welcome lower again to a few other 1/2 hour glass about the deep net, the entire advent in present day-day elegance. We are gonna see a list of Websites or maybe go to a few Web websites at the deep internet. So how will you get right of entry to the deep web? The maximum solid way. Um what you could do. Um the primary thing that I endorse you men do is in fact to get entry to the Haydn week that is kind of a Web website at the deep net in that you have were given some component first-rate.

Well you have precise lessons and you have extraordinary internet web sites so it's miles an incredible start. Just start at the deep internet. So allow's get there. So as you may see the immoderate and secret's like a Wikipedia on the s model. Deep inside the deep internet version. So as you could see you have severa schooling there. You have. Let's see that there are large volunteer monetary offerings and commercial enterprise offerings and you could find out a few bizarre stuff right here, as an example the gun marketplace. Some additionally do this stuff. Well it is a laugh to look at but I don't propose you go to those locations. I advise you have were given were given blogs in thrilling places.

You have email offerings on the deep internet that can genuinely permit you to ship emails anonymously. You additionally have social networks as you could see you have got a Facebook model of the Deep Web. You have forums. You have a whole lot of various stuff. Um, so we are going to visit one of the Web sites. Well a number of those Websites which might be interesting. The first one that I can recommend to you is the SEC mail. So I suppose it is within the electronic mail magnificence. It's this one proper right here. So it looks as if this too. It's an electronic mail company. Very truely clearly be part of up you create your uh you pleasant create your email. So all yet again they do no longer ask you to your name.

They do now not ask you for something. And which you have an email with them and you may get preserve of emails despatched and it's far all absolutely anonymous this is without a doubt cool. Um additionally properly as I said you can supply emails anonymously. Besides which you have many special subjects. Also at the deep internet I recommended you men you have got are searching out engine serps like google. One of the most famous is torch, so torch is someplace right proper right here who could have torch right there all over again you may see Web internet websites are loading very slowly as you could see you've got a search engine this is torch so we are able to as an example. Well I do now not advise you to. I do not endorse you write down matters on serps like google and yahoo however just to do that you have search engines like google proper right here as uh as you have on the clear net.

Also um you have were given got moreover the Doug the Google version of the deep internet. So we observed it in our preceding elegance. And as I stated you have got lots of nicely precise training. I propose you truly go searching. Um what else. Mm mm mm mm mm mm mm mm mm. There are some weird subjects. For instance, you can speak well with random people. It's the element proper proper proper here. So it is like a social network in which you could talk to one man or woman or you may agency chat with random humans so allow me display you what it looks as if. So you said it's like a random chat in which human beings talk with every one-of-a-type. So they said Do you've got got pretty a few weird stuff on the deep internet? The majority of it's miles quite dull and stupid however you could find interesting topics.

So as I said men the primary area which you need to visit at the same time as you're beginning out at the deep internet can be clearly the peak and weekend on the height and we're able to find out all of the categories that exist at the deep internet we actually have Websites which might be in one-of-a-kind languages. So as an example in case you need French internet sites economic Web web websites German Web net web sites you have were given got all this right proper here. It's an notable place to begin. What I endorse to you is also to go to three boards or places in which you can talk with extraordinary people. For example the blogs or boards I recommend you absolutely have a look at those um you could discover thrilling records about the deep internet as a whole and approximately the deep net they're capable to signify you some places and some links. For instance in case you want books, if you need tune.

Well you have were given got all this right proper here. If I'm not incorrect you have got had been given uh nicely you've got got got books right there. You can simply test in. You'll find out that that explains each. Each link has this creature description. Next to it um what else. I assume it's far it. Um, the identical subjects that you have at the easy on the floor net you have the identical variations of this on the deep net and you could do the precise same subjects at the deep internet that you can do at the floor net. The simplest difference is that on the Deep Web. Well it is all anonymous so that you find out it a bit more stuff. But except that it's far uh the same trouble. And as I stated the best difference is that it's miles nameless. So it sincerely is it for this Book Gates. And see you all in our subsequent elegance in which we are going to speak approximately tales.

Introduction to TAILS

All right. So once more top day sure and welcome lower back to 3 different of our Books. I want an entire creation to the Deep Web. So those glasses we are going to talk specifically approximately recollections and well for a excursion possibility and we're going to reputation on tails. So I'm going to provide an reason for exactly what tails are. I'm going to show you the Web net site of tails but we aren't gonna do the entire setup when you do not forget that this glass is surely for.

Once yet again it is very essential for you guys to have an introduction to testimonies in exactly the way it really works and what is the distinction amongst tails and that. So allow's begin all proper. So what is tails? So the precept distinction among tails and tails is that the principle difference among tails and tour is first that the excursion is like an internet browser. Well Tor is a web browser. It's a manner for you to connect to the deep internet for it. And the possibility Web internet site on-line however for tails on the other component it is the entire walking tool.

So at the same time as you set up tails you are simply installing a complete strolling system that will help you cowl your identification. And it without a doubt works with Tor. So it allows you. It allows you stay nameless. So whilst you put in tables that I stated you set up tor on tails and you may artwork with every of them. So you're a nun there virtually may be the max. Is it better than horrible? Well no longer truly because of the reality as I stated the Tails is virtually an strolling gadget. Tour is the browser higher. It may be correct almost. Yes it's far better. Why? Because while you are, you're able to use tails anywhere you need. Why? Because properly tails isn't mounted in your laptop.

Tails can be mounted on the EU is top or it is going to be set up on a DVD. So because of this that you may be part of right away to the deep net on any laptop that you want with out leaving no. Without leaving no hint. So you may leave no trace of that in case you use a computer to connect with the internet browser. This is why it is cool the use of this. Well that is still tails also. Well this is why in full-size tails it's miles a chunk better than tor except that it is quite hundreds all the ranges of protection there. There isn't always any large difference between them. Tails may be run on Windows Mac on Linux. So it is able to be run on all kinds of machines which is also very very cool.

And as I stated the principle the precept great of tails is that it can run with it is able to run right now on any laptop simply with a is high so that you make the setup handiest whilst you installation your software as quickly as and then after that you can run it on any tool that you want or a virtual machine if you men use a digital machine. And you apprehend a chunk. Well you could run it on the digital tool in addition to at the regular device. The 1/3 hassle is it's miles tough to set up. Well actually not. If you guys are inquisitive about putting in region tails I advocate you really have a look at the stairs that they've on their

Website. We're going to peer it a piece later in this elegance. So what a glance.

Well what the Website seems like and in which precisely you need to go to to put in your properly to install this app if you guys are involved to apply it all over again this Book is surely about using Tor but you could use tails due to the truth store it includes also. It's best a device. But no it's now not difficult to installation in any respect. It took me. You want a few additives Of direction you want. It takes approximately two hours of time. You additionally want eight gigabytes, a manner of as a minimum eight gigabytes. But the manner they offer an purpose of it it is very quite simple. It's quite simple to install can Thales be hacked or compromised.

Yes. If the device which you are the usage of let's imagine you're the use of a laptop. If the hardware of the computer is compromised while statistics can be compromised as nicely and the individual that has the pc well can see what you have completed on it however if the pc isn't compromised and if it is some thing safe as an example permit's anticipate you're the use of a laptop and also you need to hook up with the deep net from a espresso keep. Well nobody's going to be conscious it because of the truth well you're connecting reminiscences whether or not or now not you are key for your computer and it's miles not going to leave no hint which you related the profits to the computer. So over again that is why it is first rate to have. And the reality that you can use it to connect with the deep net from any pc is definitely in reality an super issue to have.

So allow's attempt to discover that out on the internet so I can display you men how this software program works. One less detail is open source software program software. So over again it's now not going to price you. I think it is truly free. And all over again in case you guys need to use it as an possibility to artwork you may. It's an extraordinary device. So let's have a look at what it looks as if. So the number one element you need to do is open up your Web browser. Let's say it's miles Google Chrome. You really circulate on Google and also you write down reminiscences. So proper here we flow: tails is this program proper proper right here. So as you could see you'll be in this Web internet site on line proper. That have grow to be in French and you placed it in English.

If you need to put in it you absolutely click on deploy. From this detail they'll ask you in which you want to put in it. You can set up on Windows Mac or Linux really click on on the house home windows it may have a windows and as you could see they may inform you what precisely you will need for this installation you may want. You may be key to at the least 8 gigabytes and then it's going to take about one hour. You ought to take a whole lot much less in case your net pace is faster. But besides that you do no longer genuinely do some issue. You do not want some thing extra. When the installation is completed yet again you can want to restart your laptop. And from this aspect there can be.

Why do you grade by grade how you may installation this software software software program immediately on your laptop while below you will be the important aspect that you may be able to use after that during your pc. So I stated Guys , it actually is an incredible device. Once yet again we aren't going to go extensive of this device due to the fact we are focusing great on the tour for this class. And then all over again it is clearly an creation to the deep net, it is now not a complete protection class. So we are going to maintain on with it. But in case you men want to move greater in-intensity and studies a bit greater about tells. Well I genuinely suggest you to use it because of the truth all over again it's far an outstanding way to get admission to the deep net from any form of. So it sincerely is it for this magnificence guys. And see you all in our next beauty.

Chapter 6: Crypto Currencies

All right. I'll see you guys. Welcome over again to a few different of our tremendous commands approximately the deep internet. The complete advent in extremely-modern-day beauty we're going to talk approximately deep net marketplaces and techniques that you can purchase products on the Deep Web. The first element that you men will need to realize is that the deep net or the tour browser everything that is across the deep dark internet is made to be and stay anonymous. So the manner that you are shopping for stuff in this shape of market is likewise made to be and stay nameless.

This is why you may typically well in 99 issue ninety nine percentage of the time no longer use your credit score score score card because of the fact if a market asks you for a credit score rating card you can truely be hacked. All your credit card information so in no way located any of your bank credit card or any records on any deep internet marketplaces on the ones marketplaces you may use an nameless way of purchasing the goods that you had been shopping for. Let's say you're buying a tv on the Deep Web. You are going to be nameless for this television.

How you could pay for it. Well you may bait with cryptocurrency And that is what this Book can be approximately. We are going to speak approximately crypto currencies within the subsequent elements as nicely . In the following commands we're going to talk about how to shop for you, the manner to make purchases and a way to navigate that on marketplaces. So permit's start as I said, what's a crypto forex? The first element is that it is an Internet based medium of change. What does it advise truly way that it is a manner to make purchases on the Internet or in any well, especially on the Internet.

This kind of overseas cash have emerge as very very well-known round the area. It came to life in 2009 with the bitcoin and feature end up very very famous within the years that determined because of its anonymity and the reality that you can not surely trace a transaction because all over again it's miles simply nameless and the manner that it's miles made you have public keys you have got were given non-public keys and all this creates some issue that gives anonymity to each transaction. Well it offers anonymity to each transaction because of the reality you'll in no way have a name, you may in no way have an deal with or some issue. And it's certainly truely an nameless transaction. What brings us to the second one factor is that it's no longer managed via any number one authority so how can it stay anonymous?

Well there aren't any banks that manage permit's bear in mind the bitcoin in the direction of the ETF or any kind of crypto forex. And this is an outstanding manner to keep all transactions nicely constant. It's like if you were happening permit's assume at the streets and you buy something cash allow's consider you've got been going to someone who's promoting ice cream and you buy an ice cream cash. Well no one will apprehend except the person that sold ice cream that you acquire the ice cream coins nicely which you buy an ice cream. But if you can, permit's anticipate you buy this ice cream with a credit score score card or with a credit score card or a debit card.

Well it'll depart a hint in your economic enterprise declaration. It's going to say that oh you made this transaction that day. But whilst you purchase with the crypto forex there's no bank announcement. There isn't any properly there isn't whatever it is like figuring out to buy some issue coins to a dealer which you do no longer realize and he would not understand you. So it is certainly truely anonymous and it's far in fact about numbers. You do not know the vendor. The supplier may want to not comprehend. Don't apprehend you. You transferred the cash to him. He transfers the products to you and there may be no middleman. This is the beauty of crypto foreign places cash and buying properly with using crypto forex. This is why as I stated it's miles completely anonymous and there can be.

Yeah well there are approaches to retrace transactions however all yet again it is very very difficult to trace because it's on the block chain generation. It's not it isn't always like ordinary transactions when you have let's assume what a credit score score card you have a monetary employer declaration that declares Oh that day you went to that hold and to procure a few element in this in this case nicely in the case of using crypto currencies you have got surely whole anonymously because you haven't any monetary institution declaration. Well no longer excellent greater financial institution statements you have got got got in fact no information from wherein the transaction went and who despatched you the goods or some thing it's miles like illegally like shopping for some thing coins from a vendor that doesn't recognize you and you do now not know him. So just maintain in mind it that way. All right. So the way it works for folks who are interested in how crypto foreign money works.

Once once more it is now not a crypto overseas cash Of path but just for you men to understand the manner it really works. Let's say someone requests the transaction through a dealer or change requests a transaction. Let's say you request the transaction from the vendor. So after that the request may be nicely we're going to be broke. Yes it might be too reasonably-priced to burn the network. That is ordinary. That consists of laptop structures and all the ones computer systems, permit's name them. After that this network will entire the transaction and the block chain era is how I can supply an explanation for it. It's genuinely allow's count on it blocks the way it's miles constructed. Imagine blocks one after the opportunity and also you can't change let's consider one block with out converting all the blocks in advance than him. So this creates the safety of the transaction.

And that this could now not be hacked. Well it is very very tough to hack a block chain generation considering it's miles a trendy generation. And yet again it is very very hard to retrace all the transactions because of this because of the reality as I said it's certainly properly you may see as blocks. But over again it's very complicated to apprehend. And subsequently at the prevent of all this way even as the assessments transaction is completed typically transactions that involve crypto overseas cash take region. Well , it's miles nearly real there. Well they are made inside the 2nd line.

Let's say you're making the transaction and it takes one to two seconds to finish the transaction. So over again it's miles very very steady. And in order I said there is no wealth. There is constantly a sure hazard concerned however in our case shall we say you're sitting at the slight side and you are buying stuff like tv or permit's count on computer structures or some element on the deep net. There is not any chance for you guys too nicely to have any troubles with crypto foreign exchange use.

The only hazard that I recognise it is no longer absolutely related to searching for or promoting stuff is let's assume you purchase the crypto foreign cash and keep it considering the fact that it's far very very unstable. It's not typically an superb idea to do. Because all another time the charge fluctuates loads and they are able to pass truely excessive the identical way that it may pass absolutely in fact low. So you could lose or win lots of cash however over again it's miles now not, it is not a threat that is related to purchasing for stuff at the deep web. So right now you recognize the way it simply works and you apprehend what a crypto overseas cash is. So that completes our Book and c o in our subsequent splendor in which we are gonna speak approximately how to buy crypto remote places cash and wherein to find out it.

How To Buy Cryptocurrencies

All proper. So another time hi there men and welcome back to another of our terrific Books about the deep net, the complete creation elegance. We are going to talk approximately how to shop for crypto currency. So proper now and so now you comprehend what's the crypto foreign exchange. But you need to understand the concept on the again of purchasing crypto overseas money and a way to apply it. So in the ordinary international we've coins which we will name combat cash.

When in a story this cash have become going to be saved in a monetary organization and we've were given a financial organization account then we're able to pay with our card for the crypto currencies which suits pretty a good buy the equal way. So you are taking your crypto, you buy your crypto forex and also you keep it. You need to keep it somewhere so that you can store it at once on your laptop. But it is not the motive right right here. Our aim is to have get right of entry to to it very very without problems. So we are going to use the pockets there. That is direct online so keep your crypto foreign money you'll use it properly. It can be a wallet.

As I said, it may be stored to your laptop right away. It can be a cold garage pockets this is stored on a small server or a small difficult pressure it's stored someplace at your house. But over again we're going to talk nice about wallets which can be online. So off the wall what we are going to speak approximately these days can be Coinbase and block chain dot com. Those wallets are, I count on , very , very useful and clearly smooth to use. We aren't going to swell the images that you are going to look there aren't my account. So it is truly an example of methods Social bills are a piece bit what it is and the way it works and why they are very cool to use.

So the primary one that you need to look may be the entire Coinbase. So while you get on the Web net internet web page you may ought to create your self an account. So you'll need to add your email deal with and safety diploma. It's safety due to the truth whenever which you want it that you could be part of you'll accumulate an asset mass and you may ought to enter your code. So it's far very very steady. Same thing for block chain. The way it really works you get connected whilst you get related you may see you'll have your watch list right right here. So you may have all of your cash and the price and what form of they value.

You can be in a role to buy them. Well at the same time as you add a price technique you may be in a role to buy them. So you could buy what is cool on that coin is that you should purchase them you can promote them. It's like actually actual time. So permit's expect you buy one bitcoin if the fee goes up. Well we are able to resell it. So you can make cash purchasing for and promoting, allow's recollect for example you do not flow into on the darknet and can straight away alternate your crypto currency on Coinbase this is simply clearly cool. Also at the same time as you buy your crypto foreign money it'll be stored at once proper right here. So you will be capable of see it for your portfolio for the instant as you can see I don't have any future possibilities.

And the crypto currencies that you want that you are going to shop for the most can be the Bitcoin which is the crypto forex that is the most used on the deep internet theorem also. And for the rest nicely you've got were given more narrow additionally but once more it in truth relies upon at the Web websites which you go to. Let's say a Web website on-line whilst you get the majority of the Web net web sites at the deep net will typically use bitcoin. Well let's assume some one-of-a-kind Web net web page on line a Web net site on line asks you for a particular cryptocurrency for instance do you want to sell your products but best in dash. Well you may be able to come lower again proper here and purchase your sprint cryptocurrency this is all once more truely simply cool due to the truth if for example you want to trade your crypto foreign coins you can get traded right now here.

Also in case you invite a pal at the same time as you get a small rate except that you have all of your charges proper here. So for all the crypto currencies for the immediate they have terrific 19. As you can see it varies from time to time they upload a few new crypto currencies sometimes well they did delete a number of them so yeah it's miles in reality what's cool with the coin is that as I said it's far a platform that you could use as a pockets but it is also a looking for and selling platform so shall we embrace you're someone who want a exchange war who is shopping for and promoting properly you may do it at once on Coinbase moreover if I'm now not wrong those are definitely the awesome corporation of pals is you have got loads greater of different crypto currencies proper right here.

Once the give up of the crypto currencies which might be the most used for like shall we say for getting merchandise is probably those ones proper proper here. Well the bitcoin and the ETF. And I stated probably Monday or on some web sites. But all over again it in truth depends. All proper. Next platform can be block chain dot com. So it's far sincerely the identical manner you need to shop for yours. Well the number one trouble you need to do is create your account. So you cross on block chain dot com. Once your account is created this platform is surely for buying and selling cryptocurrency. So you can have this dashboard right proper right right here. So in case you need to buy crypto you'll actually click on there you click on on purchase crypto. They will ask you first. You will not see this hassle right there.

You will need to pick out your foreign cash so that you virtually pick out out your first forex with which you want to shop for your crypto. So allow's assume I'm going to apply a British pound. So permit's expect I need to shop for 100 British pounds of crypto forex and as you could see they'll join me right right here after which we will want to fill all my non-public records after which I can purchase my cryptocurrency the usage of a credit rating card or a PayPal account. Usually it is very pretty smooth, too well to buy crypto. Once you have your crypto overseas cash the way he goes to work you can have a public key and a non-public key. And even as you want to make your transaction typically you'll only placed on your public key. You will in no way write down your private key.

And it's far tremendous, it's miles pretty simple. Let's say you need to. I do not know if you want to change properly otherwise you want to send that and deliver it to a person. So I pay attention you'll write down the overall public key of the person and you'll surely deliver them your. Well how hundreds you want to deliver crypto to this character you could ship it in coins you could supply bitcoins. This works the equal way while you purchase products on the Deep Web. Let's say you choose out a product to buy. You will most effective input your public key. Normally you do now not enter your private key due to the fact this one you hold it for you. And yeah so it is it it's far very pretty clean.

Also you may borrow crypto currencies. But all over again you want to be at a better degree. So as I stated you could purchase your crypto foreign cash there. So it is right here in case you need to change your crypto currency that's let's assume alternate bitcoin for Ethereum it is gonna be there. So you may want to have crypto distant places cash first. We want to shop for a few. And it clearly is it. Well you could alternate furthermore. Once over again our intention is virtually to buy however crypto isn't to alternate or a few aspect. So to buy a crypto overseas cash over again very quite easy structures use Coinbase block chain they don't require or I don't know how an entire lot verifications.

Very smooth to apply. You enter your call and the whole thing and you could purchase your crypto. Once you have got your crypto stored there you congratulations you have got a pockets with the crypto distant places coins. And from this 2d you are able to shop for things on the deep net so it is wherein the scorer's men right now apprehend exactly how to buy you a crypto foreign cash and wherein this crypto remote places money might be saved at this 2nd. We will see how we should purchase your merchandise at the deep internet.

Chapter 7: What Is Dark Web?

The dark internet is a hidden and encrypted segment of the net that operates past the reach of conventional search engines like google and yahoo like google and yahoo. It bureaucracy a part of the broader deep net, encompassing all on line content material cloth now not indexed for public get right of access to. What distinguishes the darkish net is its intentional emphasis on anonymity and privateness.

Users get proper of access to the darkish net the use of specialised system like Tor (The Onion Router), a network that redirects and encrypts internet net page visitors thru a sequence of volunteer-operated servers, making it tough to hint users' identities. This layered encryption is represented thru the onion metaphor, symbolizing the multiple layers of protection.

The dark net is frequently associated with illicit sports activities due to its functionality to provide clients with anonymity. Illegal marketplaces on the dark net facilitate the exchange of medicine, weapons, stolen facts, and other contraband. Cryptocurrencies, specifically Bitcoin, are commonly used for transactions, offering a in addition layer of anonymity.

However, it's far essential to observe that the dark internet isn't always entirely a hub for illegal sports activities. It additionally serves as a secure haven for people seeking out privacy in oppressive regimes, a platform for political dissidents, and a place for newshounds to talk without fear of censorship.

Despite its twin nature, the dark net remains a complex and evolving virtual panorama, elevating moral, crook, and technological questions about the stability among privateness and safety in the interconnected worldwide of the net.

The darkish internet is a clandestine and encrypted corner of the net that operates beyond the conventional obtain of stylish search engines. As a subset of the deep internet, it consists of on line content cloth fabric not indexed for public get admission to. What units the dark internet aside is its planned emphasis on privateness and anonymity. Users access it through specialised gadget like Tor (The Onion Router), which redirects and encrypts internet site site visitors via a series of servers, masking their identification and region. This intentional concealment has fostered a popularity regularly related to illicit sports activities sports, because of the fact the dark internet serves as a platform for the change of unlawful items and services, facilitated via using cryptocurrencies for transactions. However, it's essential to apprehend that the dark web isn't always absolutely a haven for cybercriminals; it additionally gives a shelter for human beings searching out privateness in oppressive regimes and a space for activists and reporters to speak without worry of censorship. The duality of the dark internet, as every a sanctuary for privacy and a shadowy realm of illicit dealings, provides layers of complexity to its function in the broader virtual panorama.

THE ACTUAL WEB

Exploring the Depths: A Comprehensive Analysis of the Deep Web Iceberg**

The internet, just like an iceberg, exhibits merely a fraction of its expansive fact on the floor. This iceberg analogy serves as a compelling framework for understanding the multi-layered components of the web worldwide, mainly because it relates to the concealed dimensions known as the Deep Web and the Dark Web.

Surface Web - The Visible Peak :

At the apex of this metaphorical iceberg lies the Surface Web. This represents the publicly handy part of the internet, resultseasily navigable through traditional search engines like google and yahoo. Everyday websites, social media systems, and widely available records comprise this acquainted territory that users interact with on a everyday foundation.

Deep Web - Hidden Depths :

Below the ground, we come across the significant Deep Web, a massive a part of the internet that stays past the acquire of conventional search engines like google. This hid realm includes proprietary databases, instructional sources, certainly one of a kind information, and different content material necessitating precise get right of entry to credentials. The Deep Web is, in essence, the submerged part of the iceberg that eludes informal statement.

Dark Web - The Submerged Realm :

The least seen and most enigmatic layer is the Dark Web. Accessible only via specialised gadget like Tor, the Dark Web operates as a haven for anonymity. It incorporates every prison sports, along with privacy-aware communications and activism, and illicit corporations, together with illegal markets and cybercriminal operations. The Dark Web, just like the hidden depths under an iceberg, stays largely concealed from the prying eyes of the not unusual internet client.

Complexity of Anonymity :

Anonymity stands as a cornerstone of the Deep Web, specifically on the Dark Web. Users, starting from privateness advocates to cybercriminals, leverage encrypted networks to difficult to apprehend their identities. The intricacy of this anonymity lies in the coexistence of legitimate privacy goals and sports that breach criminal and ethical barriers, posing a chronic challenge for regulators and regulation enforcement.

Specialized Tools :

Navigating the Deep Web necessitates the use of specialized device designed to make certain anonymity and safety. Tor, as an example, allows nameless communication via routing net website online visitors thru a chain of volunteer-operated servers. These gadget offer customers with the manner to discover hidden layers with out compromising their identification, which incorporates a technological layer to the iceberg analogy.

Legal and Ethical Considerations :

The dual nature of the Deep Web turns on crook and ethical issues. While it offers a secure haven for privateness-aware people, political activists, and whistleblowers, it additionally serves as a area in which cybercrime and illicit transactions thrive. Striking a stability a number of the proper to privateness and the want for law enforcement intervention stays a continual societal challenge.

SURFACE WEB

Surface Web : Navigating the Visible Landscape

The Surface Web, often taken into consideration the top of the large net iceberg, constitutes the visible and effortlessly handy part of the World Wide Web. This expansive digital terrain, corresponding to the bustling streets of a city, is in which customers interact in normal on line sports, effects exploring web web sites, social media systems, and a myriad of informational resources.

At its center, the Surface Web encompasses web websites and content cloth which may be listed via conventional serps which encompass Google, Bing, and Yahoo. This indexing lets in for the seamless retrieval of statistics, making it effects to be had to all and sundry with a web connection and a device. From data articles and academic belongings to e-change structures and leisure hubs, the Surface Web is the virtual area in which customers behavior searches, share facts, and participate in the worldwide alternate of thoughts.

The structure of the Surface Web is based totally completely on the concepts of accessibility and man or woman-friendliness. Websites appoint preferred protocols, inclusive of HTTP and HTTPS, permitting clients to navigate via clickable links and interact with content cloth intuitively. This open accessibility fosters a diverse on line surroundings, supporting organizations, educational establishments, media retailers, and those in sharing statistics and connecting with a global goal market.

Despite its familiarity, the Surface Web isn't without its complexities. It constantly evolves with the ever-converting panorama of the internet, driven via improvements in technology, layout, and consumer enjoy. Security worries, which consist of the implementation of HTTPS for encrypted connections, play a important feature in improving person protection and privacy at the Surface Web.

However, it's important to recognize the restrictions of the Surface Web. While it paperwork the seen part of the net iceberg, it represents high-quality a fragment of the whole digital panorama. The majority of facts, databases, and content material material live within the unseen depths of the Deep Web, inaccessible to standard search engines like google and yahoo and requiring precise credentials for get proper of entry to.

In summary, the Surface Web serves because the gateway to the net, wherein customers navigate the acquainted terrains of net sites and on line structures. Its openness and accessibility have converted the way we get right of get entry to to data, behavior business enterprise, and talk. As we hold to surf the Surface Web, it's critical to renowned its coexistence with the concealed layers of the digital realm and to understand the complicated interaction that defines our online tales.

DEEP WEB

Deep Web : Unmasking the Hidden Layers

Beneath the acquainted ground of the net lies the expansive and enigmatic realm called the Deep Web. Unlike the Surface Web, which constitutes the visible and with out trouble on hand part of the World Wide Web, the Deep Web operates beyond the advantage of traditional serps like google and yahoo. It is a hid panorama that consists of a vast part of the net, website hosting a myriad of property, databases, and content material material that stay hidden from informal on line exploration.

The Deep Web consists of content material that is not indexed via search engines like google and yahoo because of its dynamic and regularly non-public nature. This hidden expanse consists of educational databases, proprietary corporate statistics, medical statistics, and specific personal belongings. Access to the ones concealed layers frequently calls for unique credentials, making the Deep Web a repository of treasured and sensitive records.

One of the defining talents of the Deep Web is its numerous form of content material material, serving numerous capabilities past public intake. Academic institutions, studies centers, and companies employ this hidden area to store and change information that isn't intended for public view. It additionally consists of personal networks and intranets, in addition contributing to the vastness of the Deep Web.

Navigating the Deep Web necessitates a departure from traditional are looking for techniques. Specialized tools and protocols, at the side of Tor (The Onion Router), I2P (Invisible Internet Project), or Freenet, are hired to get proper of entry to the ones concealed layers. These tools provide a diploma of anonymity and encryption, contributing to the privacy and safety of customers exploring the hidden depths.

It's crucial to distinguish the Deep Web from the Dark Web, because the former consists of a broader spectrum of content fabric, each criminal and illicit. While the Deep Web hosts precious and valid assets, the Dark Web, a subset of the Deep Web, is thought for clandestine sports activities facilitated thru encrypted networks. These activities range from privacy-centered communications to illicit transactions and cybercrime.

In summary, the Deep Web represents the unseen majority of the internet, concealing a wealth of data that extends beyond the limits of public visibility. As generation advances and our digital landscape evolves, know-how the complexities and capability of the Deep Web will become increasingly crucial, emphasizing the want for responsible exploration and a nuanced method to the multifaceted layers that define our on-line enjoy.

Chapter 8: Navigating The Shadows Of Cyberspace

The Dark Web, a hidden realm beneath the floor of the internet, exists as a mysterious and enigmatic location that stands in stark assessment to the visible and listed net. Often perceived as a virtual underworld, the Dark Web operates on encrypted networks, intentionally hid from traditional search engines like google, and handy high-quality through specialised equipment like Tor (The Onion Router).

The Cryptic Landscape :

The Dark Web is a subset of the wider deep net, which encompasses all factors of the net no longer indexed with the resource of conventional serps like google and yahoo. What devices the Dark Web apart is its emphasis on anonymity and privateness. Users seeking safe haven from the prying eyes of mainstream online systems have interaction in encrypted communication and transactions, their identities shielded with the useful resource of layers of obfuscation.

Accessing the Abyss :

Navigating the Dark Web calls for a departure from the familiar click on on-and-are in search of method of the floor net. Specialized machine like Tor anonymize customers via way of routing their internet visitors via a series of volunteer-operated servers, making it hard to hint their foundation. The time period "onion" in Tor displays the layers of encryption that encapsulate character information, growing a virtual veil that obscures identities.

Duality of Purpose :

The Dark Web, however, isn't always a monolithic entity defined certainly by way of manner of way of crook sports activities. While it undeniably hosts illicit transactions and cybercrime, it additionally serves as a shelter for privacy-conscious human beings, political dissidents, and journalists going for walks in environments adverse to loose speech. Encryption gear that ensure anonymity on the Dark Web can also empower those seeking out protection from surveillance.

Illicit Activities :

The Dark Web's notoriety regularly stems from its affiliation with illegal trade and cybercrime. Underground marketplaces facilitate the alternate of unlawful goods and services, together with drugs, guns, stolen data, and hacking system. Cryptocurrencies, especially Bitcoin, are often used for transactions, supplying a layer of monetary anonymity.

Challenges for Law Enforcement :

The decentralized and anonymized nature of the Dark Web provides huge demanding situations for law enforcement. Investigating and prosecuting crook sports activities activities on this digital abyss require specialised know-how, technological understanding, and worldwide collaboration. The very capabilities that manipulate to pay for privateness to valid customers additionally offer a haven for those engaged in illicit agencies.

Ethical Considerations :

The moral landscape of the Dark Web is complex and nuanced. It will increase crucial questions on the steadiness between privacy rights and the want of enforcing criminal suggestions to guard society. The very anonymity that safeguards unfastened speech and privateness additionally offers cover for crook sports, growing a perpetual tension among individual freedoms and collective protection.

Technological Evolution :

As era advances, the Dark Web evolves in reaction to countermeasures taken via government. The cat-and-mouse recreation some of the ones searching out privateness and people aiming to put into impact the regulation drives persistent innovation in each encryption era and investigative strategies.

The Future of the Dark Web :

The destiny of the Dark Web stays unsure, normal through ongoing technological tendencies, crook responses, and societal attitudes in the direction of privacy and safety. As societies grapple with the demanding conditions posed with the resource of this hid digital panorama, a nuanced knowledge of its complexities will become crucial.

In quit, the Dark Web stands as a paradoxical area, concurrently harboring criminal sports activities and imparting secure haven for the ones looking for privateness in an interconnected global.

HOW DOES THE DARK WEB WORK

Unveiling the Mechanics of the Dark Web: An In-Depth Exploration

The Dark Web, shrouded in layers of encryption and deliberate obscurity, operates as a clandestine realm within the broader landscape of the net. To understand how the Dark Web works, one ought to delve into its particular shape, the technology that underpin it, and the myriad sports activities that spread in its encrypted corridors.

1. Anonymity via Encryption:

Central to the functioning of the Dark Web is the emphasis on anonymity. Users gaining access to the Dark Web deploy specialised system like Tor (The Onion Router) to obfuscate their identity. Tor operates through bouncing customers' internet web site site visitors thru a sequence of volunteer- operated servers, encrypting the facts at every step. This layered encryption, represented via manner of the metaphor of an onion, guarantees that the supply of the individual's interest remains concealed.

2. Tor Network - The Backbone of Anonymity:

Tor, the cornerstone of Dark Web anonymity, capabilities as a decentralized network of volunteer-operated servers. Users hook up with the Tor community via a browser configured to direction their net visitors via this series of servers. The encrypted records packets are transmitted randomly thru multiple nodes, making it pretty tough to hint the foundation of the communique. The use of Tor successfully anonymizes each the sender and the receiver of the data.

three. .Onion Domains - Concealing Identities:

Websites on the Dark Web hire ".Onion" domain names, a one in every of a type function that differentiates them from conventional internet addresses. These domains are not indexed through ordinary search engines like google and yahoo, contributing to the concealment of Dark Web web sites. The difficult combination of Tor and .Onion domain names creates a closed surroundings wherein customers can have interaction in sports a ways from the scrutiny of the surface internet.

four. Decentralization and Resilience:

The Dark Web operates on standards of decentralization and resilience. Unlike the centralized shape of the floor internet, that is primarily based on particular servers and records centers, the Dark Web's allotted nature makes it greater resistant to censorship and shutdowns. If one node is compromised, others can seamlessly take its place, making sure continuity.

five. Illicit Marketplaces and Cryptocurrencies:

One of the exquisite abilties of the Dark Web is the life of underground marketplaces facilitating the change of illegal goods and offerings. Cryptocurrencies, in particular Bitcoin, are the favored medium of exchange. The decentralized and pseudonymous nature of cryptocurrencies aligns with the ethos of the Dark Web, supplying an additional layer of economic anonymity for customers engaged in transactions.

6. Privacy Tools and Encrypted Communication:

In addition to Tor, privacy gear in conjunction with encrypted messaging offerings play a important characteristic in securing communications on the Dark Web. PGP (Pretty Good Privacy) encryption, for instance, permits customers to deliver and collect messages securely, further safeguarding their identities and the content material cloth of their communications.

7. Challenges for Law Enforcement:

The very features that offer privacy and safety at the Dark Web pose extensive annoying situations for law enforcement. Investigations into crook activities end up complex, requiring specialized expertise and global cooperation. While regulation enforcement agencies have evolved strategies to cope with cybercrime, the cat-and-mouse sport continues because of the truth the Dark Web adapts to countermeasures.

eight. The Duality of the Dark Web:

It's vital to understand the duality of the Dark Web. While it gives a haven for the ones looking for privacy in oppressive regimes, a platform for political dissidents, and an area for reporters to speak without fear of censorship, it also harbors crook sports activities sports. This duality underscores the complexity of the Dark Web as a virtual location.

In stop, the Dark Web operates as a very particular and difficult virtual environment, leveraging encryption, decentralized networks, and anonymity gear to create a area that is every a refuge for privacy and a shadowy realm of illicit dealings. Understanding its mechanics calls for navigating the nuanced interaction of era, privacy, and the ethical troubles that outline this concealed corner of the net.

Chapter 9: Ways To Access The Dark Web

Navigating the Shadows: Methods of Accessing the Dark Web

Accessing the Dark Web, a realm intentionally concealed from contemporary internet browsers and search engines like google and yahoo like google and yahoo like google and yahoo, requires specialized system and a nuanced expertise of its particular architecture. While the Dark Web harbors every legitimate and illicit sports, exploring it needs a commitment to privacy and protection. In this exploration, we're capable of delve into the primary strategies employed to access the Dark Web and the tools that facilitate this adventure into the virtual shadows.

1. Tor Browser: The Gateway to Anonymity:

The maximum not unusual and broadly diagnosed technique of gaining access to the Dark Web is thru the Tor Browser. Tor, or The Onion Router, is a decentralized network that directs net traffic thru a chain of volunteer-operated servers, encrypting the facts at every step. The Tor Browser, constructed upon the Firefox browser, is configured to navigate this community, allowing clients to get right of entry to web web sites with .Onion domains—specific addresses unique to the Tor network. Users can down load the Tor Browser absolutely free, supplying a as a substitute person-fantastic get right of access to problem into the encrypted corridors of the Dark Web

2. Specialized Search Engines:

Dark Web users frequently rely upon specialised serps like google designed to index .Onion internet web sites. These serps like google and yahoo like google and yahoo characteristic in the Tor community and allow customers to find out internet web sites and assets hidden from traditional serps like google and yahoo like google. Notable examples encompass "DuckDuckGo" and "NotEvil," which prioritize consumer privacy and anonymity in their are looking for functionalities. These engines serve as gateways to the various content material fabric available at the Dark Web, beginning from forums and blogs to marketplaces and verbal exchange structures.

three. I2P (Invisible Internet Project):

An opportunity to Tor, I2P is any other nameless community designed to shield clients' identities and sports. While Tor is commonly targeted on providing nameless get right of entry to to the everyday net, I2P is an included community providing each anonymity and privacy. I2P employs a decentralized peer-to-peer version, routing visitors through a community of volunteer- operated nodes. Users can get admission to I2P via the I2P router, letting them discover .I2p domain names and engage in anonymous communication.

four. Freenet:

Freenet is but each exclusive decentralized and peer-to-peer network that lets in customers to get right of entry to statistics with a focal point on privateness and censorship resistance. Unlike Tor and I2P, Freenet is designed for the strong distribution of static content material fabric. Users can put up and get right of entry to content anonymously, and Freenet employs a unique key-based retrieval gadget. It's vital to note that Freenet, whilst a part of the broader Dark Web surroundings, has a first-rate structure and use case as compared to Tor and I2P.

five. Virtual Private Networks (VPNs):

While now not a direct gateway to the Dark Web, Virtual Private Networks (VPNs) can be used on the aspect of Tor or different anonymity networks to decorate privateness. A VPN encrypts a customer's net connection, such as a further layer of safety and obfuscating the individual's IP cope with. This more step can similarly difficult to understand the man or woman's identity, especially at the same time as gaining access to the Dark Web.

6. Secure Operating Systems:

Employing a normal jogging system designed for privateness and anonymity is important whilst getting access to the Dark Web. Tails, as an example, is a stay operating tool that customers can begin on nearly any pc from a USB stick or a DVD. Tails routes all net site visitors thru the Tor network thru default, making sure a steady and personal environment for gaining access to the Dark Web.

7. Caution and Digital Hygiene:

Regardless of the technique selected to get entry to the Dark Web, caution and virtual hygiene are paramount. Users have to be vigilant about the content material fabric they get proper of get right of entry to to, avoid clicking on unverified hyperlinks, and be aware about functionality protection risks. Engaging in stable practices, inclusive of often updating software program application software and maintaining anonymity fine practices, is important to navigating the Dark Web correctly.

Accessing the Dark Web wishes a planned and informed technique, considering the privateness and protection implications inherent in its encrypted landscape. Whether via the Tor Browser, possibility networks like I2P and Freenet, or more privateness equipment, customers need to weigh the blessings of anonymity closer to the capacity risks related to exploring the shadows of the digital realm. Understanding the system and strategies available ensures a greater knowledgeable and robust exploration of the Dark Web's multifaceted terrain.

TOR BROWSER

Tor Browser: Navigating the Web Anonymously

In the labyrinth of the internet, in which data flows freely but privacy often seems like concept, the Tor Browser emerges as a beacon of anonymity and virtual liberation. Short for The Onion Router, Tor is not most effective a browser; it's a gateway to a realm in which user identities are veiled, and online sports end up obscured beneath layers of encryption.

Decentralized Architecture:

At the heart of Tor Browser's innovation lies its decentralized structure. Unlike conventional browsers, which be part of customers at once to web sites, Tor routes net web page traffic via a series of volunteer-operated servers. This network of nodes encrypts and anonymizes the records at each step, growing a dynamic and problematic net of privacy. The metaphorical "onion" layers constitute the encrypted pathways thru which purchaser facts travels, defensive it from prying eyes.

Onion Routing:

The essence of Tor's safety lies in its onion routing approach. As facts traverses thru severa relays in the Tor network, every layer of encryption is peeled away, revealing most effective the facts vital for the subsequent hop. The very last relay by myself is aware about the destination, successfully concealing each the man or woman's identification and their browsing sports. This multilayered approach to anonymity sets Tor apart within the realm of online privateness.

Access to .Onion Domains:

Tor Browser opens the door to a parallel universe on the internet—the Dark Web—via way of permitting get right of entry to to net websites with .Onion domains. These domain names, particular to the Tor network, house a spectrum of content material material, from boards and blogs to marketplaces and conversation systems. These hid corners of the internet remain invisible to standard search engines like google and yahoo like google, presenting clients a secluded area for each legitimate and illicit sports activities.

Privacy and Security Features:

Beyond its anonymity middle, Tor Browser consists of abilities committed to person privacy and safety. The "NoScript" extension, for example, blocks probably unstable scripts on web sites, fortifying defenses in competition to malicious sports. Moreover, Tor Browser enforces HTTPS connections whenever viable, making sure that information in transit stays encrypted, safeguarding customers in competition to eavesdropping and surveillance.

User-Friendly Interface:

Despite its advanced privacy functions, Tor Browser prioritizes a purchaser-best interface. Its design mirrors that of conventional browsers, imparting a seamless transition for clients acquainted with mainstream internet navigation. The intentional person-centric approach interests to make the device to be had to a broader target audience, encouraging adoption and selling privateness-aware surfing practices.

Open-Source Collaboration:

Tor Browser prospers on the thoughts of transparency and open-deliver collaboration. The task isn't always constrained within the partitions of a company entity; instead, it invites scrutiny and contributions from developers and safety professionals worldwide. This communal attempt ensures that the software program software remains sturdy, adapting to growing threats and evolving along the ever-changing panorama of on line privacy.

Challenges and Ethical Considerations:

Yet, the Tor Browser does no longer tread with out stressful conditions. Its decentralized nature, while promoting privateness, can result in slower surfing speeds compared to traditional opposite numbers. Moreover, troubles have been raised approximately the capacity misuse of the Tor community for illicit sports at the Dark Web. Balancing the virtues of anonymity with the moral obligation to cope with such traumatic conditions stays an ongoing undertaking.

In a digital technology where privateness is frequently sacrificed within the call of consolation, Tor Browser stands as a testomony to the possibility of a greater private and steady on-line revel in. It is a device that empowers clients to discover the vastness of the internet without compromising their identification. As we navigate the complexities of the internet international, Tor Browser invites us to rethink our method to virtual privateness and, in doing so, takes us on a adventure thru the hard layers of the internet's onion.

Chapter 10: The World Of Cyber Crime

Landscape of Cyber crime

In the interconnected international of the twenty first century, the digital landscape is a double-edged sword—a realm of innovation, collaboration, and boundless information, however moreover a breeding ground for a nefarious entity known as cybercrime. This virtual menace, born from the marriage of generation and criminal cause, poses profound demanding situations to humans, agencies, and global locations alike.

The Digital Epoch:

As societies globally transition into an technology dominated via virtual technology, the scope and impact of cybercrime have reached super ranges. Cybercrime includes a massive spectrum of illicit sports dedicated in the virtual realm, leveraging vulnerabilities in era to compromise structures, make the most facts, and disrupt the cloth of our more and more digitized lives.

Diverse Manifestations:

The landscape of cybercrime is multifaceted, encompassing a various array of crook sports. From the robbery of private records and monetary fraud to fashionable-day cyber- espionage and ransomware attacks, cybercriminals lease ever-evolving techniques to make the most vulnerabilities in networks, software program application software, and human behavior. The anonymity afforded thru the virtual realm gives a veil for crook actors, permitting them to characteristic with a diploma of impunity.

Global Reach and Impact:

Unlike traditional styles of crime restrained through the use of geographic boundaries, cybercrime transcends borders effects. Criminal actors, often prepared into ultra-modern networks, can execute assaults from far off places, making attribution and prosecution hard. The global reach of cybercrime poses a huge chance to the monetary stability, countrywide safety, and privacy of people and organizations global.

Targets and Victims:

No entity is proof in opposition to the tentacles of cybercrime. Governments, agencies, small corporations, and people—all find out themselves capability desires. The motives within the again of cybercrime range, ranging from financial gain and company espionage to political activism and sheer malicious motive. As the virtual surroundings continues to extend, so does the assault ground for cybercriminals, perpetuating an by no means-finishing cat-and-mouse pastime amongst perpetrators and defenders.

Evolving Tactics and Techniques:

Cybercriminals are agile and adaptive, continuously evolving their strategies to take benefit of rising technology and vulnerabilities. Social engineering, phishing assaults, malware, and zero-day exploits are just a few examples of the machine inside the cybercriminal arsenal. The rapid pace of technological innovation similarly complicates the challenge of securing digital infrastructure, requiring constant vigilance and proactive defense measures.

The Underground Economy:

In the shadowy corners of the net, a thriving underground economic system helps and fuels cybercrime. Illicit marketplaces provide a marketplace for stolen statistics, hacking device, and malware, even as anonymous cryptocurrencies facilitate untraceable economic transactions. This virtual underworld offers a haven for cybercriminals to collaborate, percentage data, and monetize their sports activities with relative ease.

The Human Element:

While generation is a key enabler of cybercrime, the human detail remains a essential detail. Social engineering assaults, which manipulate people into divulging touchy facts, are a testament to the mental techniques employed by way of cybercriminals. Enhancing cybersecurity requires no longer best technological defenses however moreover a focus on elevating focus and fostering a culture of virtual resilience.

The Road Ahead:

As we navigate the complexities of the virtual age, the war against cybercrime needs collaboration, innovation, and a holistic technique to protection. Governments, corporations, and individuals need to art work collectively to offer a lift to digital defenses, enact robust cybersecurity guidelines, and promote a lifestyle of cyber reputation. Only through a united the the front can we choice to navigate the evolving landscape of cybercrime and steady a digital destiny that prioritizes privateness, integrity, and do not forget.

DARK WEB CRIMES

Crimes Unveiled

The Dark Web, a clandestine realm available simplest via specialized equipment like Tor, isn't always excellent a sanctuary for privacy advocates. It harbors a shadowy underworld in which illicit sports activities sports flourish inside the recesses of encrypted networks. From illegal trade to cyber battle, the crimes dedicated at the Dark Web paint a complex portrait of the virtual underworld.

1. Illicit Drug Trade:

One of the most notorious activities thriving on the Dark Web is the unlawful drug change. Online marketplaces strolling with the anonymity furnished via cryptocurrencies facilitate the sale and distribution of a big range of narcotics. From marijuana and cocaine to artificial capsules, clients and sellers interact in transactions covered in the direction of traditional law enforcement.

2. Weapons Trafficking:

The Dark Web serves as a market for the ones searching out to accumulate illegal weapons. Firearms, explosives, and first rate contraband may be procured through hidden on line structures. Cryptocurrencies make certain a degree of financial anonymity, making it hard for government to trace the transactions involved in this underground arms change.

three. Cybercrime Services:

In the digital shadows, a market for cybercrime offerings prospers. Hackers-for-hire, malware developers, and distinctive cybercriminals provide their facts for a charge. This virtual black marketplace offers a platform for orchestrating assaults, stealing touchy data, and compromising digital protection.

4. Stolen Data Markets:

The Dark Web is a hub for stolen data markets in which private statistics, credit card statistics, and login credentials are provided and bought. Data breaches from essential companies regularly discover their manner onto those systems, allowing criminals to make the maximum the stolen records for identity robbery, financial fraud, and unique malicious sports.

5. Fraudulent Documents:

Forged passports, driving force's licenses, and different identity documents are resultseasily available on the Dark Web. Criminals use those solid files for numerous unlawful features, together with identification robbery, human trafficking, and facilitating one in every of a type crook businesses.

6. Hitmen for Hire:

While the legitimacy of such offerings is questionable, claims of hitmen-for-lease services have surfaced at the Dark Web. These purported offerings provide to carry out crook acts, collectively with violence and assassination, for a charge. It stays a subject of dialogue whether the ones services are actual or mere scams preying on darkish desires.

7. Child Exploitation:

The Dark Web has regrettably turn out to be a platform for heinous crimes, including the distribution and trade of infant exploitation fabric. Pedophilic groups function within the hidden corners of encrypted networks, making it hard for regulation enforcement to tune and dismantle the ones networks.

eight. Insider Trading and Financial Crimes:

Insider searching for and selling hints, stolen economic statistics, and data on marketplace manipulations are regularly traded at the Dark Web. Criminals have interaction in economic crimes with the reason of making illegal earnings, taking advantage of the relative anonymity furnished via cryptocurrencies and encrypted verbal exchange channels.

nine. Extortion and Ransomware:

Extortionists and ransomware operators leverage the Dark Web to talk with their patients and call for bills. Ransomware, a shape of malicious software program program that encrypts a victim's files, regularly needs fee in cryptocurrencies to provide the decryption key. The Dark Web gives a constant space for the ones criminal transactions.

10. Human Trafficking:

While discussions and commercials related to human trafficking may not immediately result in transactions, the Dark Web performs a feature in facilitating this abhorrent crime. From recruitment to the change of facts, the Dark Web offers a platform for human traffickers to function discreetly.

The crimes taking area at the Dark Web display off the complexity of this hidden virtual landscape. While the privacy afforded with the aid of way of manner of encryption tools can empower human beings attempting to find freedom from surveillance, it also creates a haven for human beings with nefarious cause. Combatting crimes on the Dark Web calls for a multifaceted technique, regarding technological innovation, international collaboration, and a self-discipline to upholding the rule of thumb of law inside the virtual age.

PREVENTIONS AND SAFETY MEASURES

Strategies for Preventing Dark Web Crimes

The clandestine global of the Dark Web, with its veil of anonymity and encrypted transactions, poses widespread challenges for law enforcement and those alike. However, proactive measures can be taken to save you and mitigate the dangers related to Dark Web crimes. From bolstering cybersecurity to fostering digital literacy, a multifaceted technique is essential in navigating the shadows securely.

1. Cybersecurity Fortification:

Strengthening cybersecurity is paramount in preventing Dark Web crimes. Individuals and corporations must rent sturdy safety features, which includes firewalls, antivirus software program software, and intrusion detection structures. Regular software software software updates and patches are vital to deal with vulnerabilities that cybercriminals would possibly take benefit of. Employing superior encryption protocols for touchy facts adds an additional layer of safety toward unauthorized get admission to.

2. Dark Web Monitoring:

Actively monitoring the Dark Web for mentions of an organisation or individual can be a preemptive approach. Security organizations can hire specialized system and services that check the Dark Web for compromised credentials, leaked statistics, or discussions associated with ability cyber threats. Early detection permits for quick responses and the implementation of safety abilities to save you functionality breaches.

three. User Education and Digital Literacy:

Promoting customer schooling and virtual literacy is a robust weapon toward Dark Web crimes. Individuals must be aware about the dangers related to the Dark Web, together with phishing tries, social engineering methods, and the capacity consequences of wearing out illicit on line sports. Training programs on spotting and heading off capability threats can empower users to navigate the digital landscape competently.

four. Two-Factor Authentication (2FA) and Strong Passwords:

Implementing -detail authentication (2FA) affords a further layer of safety, making it appreciably extra hard for unauthorized customers to get proper of access to debts, despite the fact that login credentials

are compromised. Encouraging the use of robust, specific passwords and regularly updating them reduces the risk of a fulfillment brute force assaults on man or woman debts.

5. Collaboration and Information Sharing:

Collaboration among law enforcement organizations, cybersecurity specialists, and personal agencies is essential for stopping and preventing Dark Web crimes. Information sharing about rising threats, attack styles, and the cutting-edge cybercriminal methodologies complements collective resilience. Public-personal partnerships can facilitate a coordinated response to the dynamic disturbing conditions posed through using the Dark Web.

6. Legislative Measures and International Cooperation:

Governments play a pivotal position in preventing Dark Web crimes through regulation and worldwide cooperation. Strengthening cybercrime prison recommendations, updating regulation to deal with growing threats, and fostering collaboration among nations are crucial components of a complete technique. International treaties and agreements can facilitate coordinated efforts to combat cybercrime on a international scale.

7. Encryption and Anonymity for Legitimate Users:

While the Dark Web is often related to illicit sports, it also serves as a safe haven for humans seeking out privacy and protection from surveillance. Encouraging the responsible use of encryption equipment and anonymizing offerings for valid functions, which include protective freedom of speech or permitting consistent conversation in oppressive regimes, enables keep the incredible elements of privacy.

8. Ethical Hacking and Security Audits:

Conducting ethical hacking and protection audits is a proactive technique to identifying and patching vulnerabilities. Organizations can rent moral hackers to simulate cyberattacks and perceive weaknesses in their structures. Regular safety audits help ensure that cybersecurity measures are up to date and effective in safeguarding in opposition to capability Dark Web threats.

9. Continuous Monitoring and Incident Response:

Continuous tracking of networks and structures is crucial for early detection of protection incidents. Establishing a robust incident reaction plan guarantees a quick and coordinated response to a ability breach. This includes predefined protocols, verbal exchange techniques, and a set knowledgeable to address and mitigate the impact of protection incidents correctly.

10. Responsible Use of Cryptocurrencies:

Given that cryptocurrencies are regularly the favored medium of change at the Dark Web, selling responsible use of those digital belongings is crucial. Encouraging adherence to jail and moral necessities in cryptocurrency transactions and implementing regulatory measures can lower the economic additives of Dark Web crimes.

In end, fighting Dark Web crimes calls for an entire and dynamic method that combines technological measures, schooling, international cooperation, and legislative responsibilities. As the virtual landscape keeps to comply, proactive strategies can be essential in navigating the shadows securely and minimizing the risks related to the clandestine sports activities of the Dark Web.